Lecture Notes in Mathematics

Volume 2382

This series reports on new developments in all areas of mathematics and their applications - quickly, informally and at a high level. Mathematical texts analysing new developments in modelling and numerical simulation are welcome. The type of material considered for publication includes:

1. Research monographs
2. Lectures on a new field or presentations of a new angle in a classical field
3. Summer schools and intensive courses on topics of current research.

Texts which are out of print but still in demand may also be considered if they fall within these categories. The timeliness of a manuscript is sometimes more important than its form, which may be preliminary or tentative. Please visit the LNM Editorial Policy (https://drive.google.com/file/d/1MOg4TbwOSokRnFJ3ZR3ciEeKs9hOnNX_/view?usp=sharing)

Titles from this series are indexed by Scopus, Web of Science, Mathematical Reviews, and zbMATH.

Anna Weller

Numerical Methods for Metric Graphs

Eigenvalue Problems and Parabolic Partial Differential Equations

Springer

Anna Weller
Department of Mathematics and Computer Science
University of Cologne
Cologne, Germany

ISSN 0075-8434 ISSN 1617-9692 (electronic)
Lecture Notes in Mathematics
ISBN 978-3-032-05010-6 ISBN 978-3-032-05011-3 (eBook)
https://doi.org/10.1007/978-3-032-05011-3

This work was supported by Excellence Initiative of the University of Cologne

This Springer imprint is published by the registered company Springer Nature Switzerland AG
The registered company address is: Gewerbestrasse 11, 6330 Cham, Switzerland

Preface

My interest in diffusion problems on metric graphs was motivated by the interdisciplinary research project "Neurodegeneration Forecasting - A Computational Brainsphere Model for the Simulation of Alzheimer's Disease" which I was fortunate to be a part of. With the goal of modeling protein propagation on networks in mind, we arrived at the formulation of parabolic differential equations on metric graphs. Since many peers showed interest in the numerical solution of such problems, my work ultimately went far beyond the practical simulation for the Alzheimer's disease use case. After all, it was my lecture on "Numerical Methods for Partial Differential Equations on Metric Graphs" and the interest of the students that motivated me to design this work in a largely self-contained scope and with many detailed practical examples. As this book origins from my research on diffusion-type problems on network-like structures during my PhD studies, it is based on my dissertation (University of Cologne, 2024 [1]) and has been modified for this publication only in some aspects, mainly in terms of editing and layout.

The purpose of this book is the systematical numerical study and analysis of related problems, in particular parabolic partial differential equations and eigenvalue problems posed on metric graphs. It contains a largely self-contained introduction to all related theoretical content, such as metric graphs, quantum graphs, well-posedness as well as rigorous derivations of the proposed numerical methods. The book is designed for researchers interested in practical aspects and numerical investigation of metric or quantum graphs and allows straightforward application of all derived methods using the accompanying julia package `MeGraPDE.jl`. Due to its self-contained nature, it is also well suited for graduate students interested in PDEs and provides plenty of material for teaching.

Cologne, Germany
June 2025

Anna Weller

References

1. Weller, A.: Numerical methods for partial differential equations on metric graphs, PhD thesis, Universität zu Köln (2024), https://kups.ub.uni-koeln.de/73182/

Acknowledgments I express my gratitude for the funding of my position as part of the project "Neurodegeneration Forecasting - A Computational Brainsphere Model for the Simulation of Alzheimer's Disease," which was supported by the Excellence Initiative of the University of Cologne as well as to the PIs Prof. Dr. Angela Kunoth, Prof. Dr. Yaping Shao, Prof. Dr. Alexander Drzezga, and also Prof. Dr. Thilo van Eimeren. I would particularly like to thank Prof. Kunoth for giving me the opportunity to participate in this project as well as for the continuous contextual support I received from Prof. Kunoth and for the constant motivation to advance my research and teaching projects on PDEs on graphs. Special thanks are also directed to Dr. Gérard Bischof for his excellent professional mentoring within the neuro-project. In addition, I am grateful for the cooperation with Dr. Merle Hönig and acknowledge that part of the data used in this work were obtained from the Alzheimer's Disease Neuroimaging Initiative (ADNI) database (adni.loni.usc.edu).

I further acknowledge that part of my research was possible thanks to a Fulbright-funded visiting research stay at Brown University. At Brown University, the excellent mentoring of Prof. Dr. Mark Ainsworth was one of the most important foundations for the development of the spectral Galerkin method presented in this work. I also appreciate the financial support from HYPATIA.Science and the GMA of the University of Cologne to support a visiting research stay at SNS Pisa with Prof. Dr. Michele Benzi, and I would like to thank Prof. Dr. Benzi for the fruitful discussions and his extremely helpful comments on this work. In this context, I also express my appreciation to Prof. Dr. Claudio Canuto for his interest in my work and his valuable feedback and comments. I would also like to thank Dr. Leonardo Robol for pointers to the literature on NEPs.

Many students I met during my time at the University of Cologne were involved in driving my work on PDEs on metric graphs forward. First and foremost Chong-Son Dröge, my first mentee as part of a program funded by HYPATIA.Science. I would also like to thank the other students who have accompanied me over the years: Max Brockmann, for his comments on this work, Lukas Schmitz and David Stiller, who worked on my favorite topic, the calculation of quantum graph spectra. And finally, all the students who actively participated in my lecture on numerical methods for PDEs on metric graphs.

Declarations

Competing Interests The author has no conflicts of interest to declare that are relevant to the content of this book.

Ethics Approval This book includes experiments with MRI data obtained at the Research Center Juelich, Germany. The ethics committee of the medical faculty of the University of Cologne approved the study. Written informed consent was obtained from all participants before the study following the Declaration of Helsinki.

Contents

Notation

Graphs and Differential Equations

Combinatorial Graphs

$\mathcal{G} = (\mathcal{V}, \mathcal{E})$	simple (undirected) combinatorial graph with vertex set $\mathcal{V}$ and edge set $\mathcal{E}$
n, m	number of vertices $n := \lvert\mathcal{V}\rvert$ and edges $m := \lvert\mathcal{E}\rvert$
v, v_i	vertex in $\mathcal{V} = \{v_1, \ldots, v_n\}$
e, e_j	edge in $\mathcal{E} = \{e_1, \ldots, e_m\}$
$e_{ij} = (v_i, v_j)$	edge between v_i and v_j
$v_i \sim v_j$	v_i and v_j are adjacent
$\mathcal{N}(v)$	set of neighbors of v
$\deg(v)$	degree of vertex v
$\mathcal{E}_v$	set of incident edges of v
$\mathrm{dist}(v_i, v_j)$	distance of v_i and v_j
$b, \bar{b}$	directed bond $b = (v_i, v_j)$ with reversal $\bar{b} = (v_j, v_i)$
$o(e), t(e)$	origin and terminal vertex of an oriented edge e
$\mathcal{E}_v^{\mathrm{out}}, \mathcal{E}_v^{\mathrm{in}}$	set of outgoing and incoming edges of v

Graph Matrices

$\mathbf{A}$	adjacency matrix
$\mathbf{D}$	degree matrix
$\mathbf{L}$	graph Laplacian matrix
$\mathbf{N}$	incidence matrix
$\mathcal{L}$	normalized graph Laplacian matrix
$\Delta_{\mathcal{G}}$	harmonic graph Laplacian matrix of $\mathcal{G}$
$\sigma(\Delta_{\mathcal{G}})$	spectrum of $\Delta_{\mathcal{G}}$
(μ, Φ)	eigenvalue and eigenvector of $\Delta_{\mathcal{G}}$
Υ	eigenvector of $\mathcal{L}$

Metric Graphs

ℓ_e	length of edge e
$\boldsymbol{\ell}$	vector containing the edge lengths of Γ, i.e. $\boldsymbol{\ell} = (\ell_{e_1}, \ldots, \ell_{e_m})^T$
Γ	metric graph
vol_Γ	volume of Γ defined by $\mathrm{vol}_\Gamma := \sum_{e\in\mathcal{E}} \ell_e$
ℓ	uniform edge length of equilateral graph
$\mathbf{u} : \mathcal{V} \to \mathbb{R}$	function on combinatorial graph, $\mathbf{u} \in \mathbb{R}^n$
$u : \Gamma \to \mathbb{R}$	function on metric graph, collection $\{u_e\}_{e\in\mathcal{E}}$ of functions u_e
u_e	function $u_e : [0, \ell_e] \to \mathbb{R}$ defined on parametrization of e

Function Spaces on Graphs

$C(\Gamma)$	space of continuous functions on Γ
$\mathbf{u}_\mathcal{V}$	restriction of u to vertices $\mathcal{V}$
$(u_e, w_e)_e$	inner product on edge e defined by $(u_e, w_e)_e := \int_0^{\ell_e} u_e(x) w_e(x) \mathrm{d}x$
$\|u_e\|_{L^2(e)}$	L^2-norm on $e = [0, \ell_e]$ defined by $\|u_e\|^2_{L^2(e)} := (u_e, u_e)_e$
$L^2(e)$	space of square integrable, measurable functions on $e = [0, \ell_e]$
$(u, w)_\Gamma$	inner product on Γ defined by $(u, w)_\Gamma := \sum_{e\in\mathcal{E}} (u_e, w_e)_e$
$\|u\|_\Gamma$	L^2-norm on Γ defined by $\|u\|^2_\Gamma := \sum_{e\in\mathcal{E}} \|u_e\|^2_{L^2(e)}$
$L^2(\Gamma)$	space of square integrable, measurable functions on Γ
$\|u_e\|_{H^1(e)}$	H^1-norm on e defined by $\|u_e\|^2_{H^1(e)} := \|u_e\|^2_{L^2(e)} + \|\frac{\mathrm{d}u_e}{\mathrm{d}x}\|^2_{L^2(e)}$
$\|u\|_{H^1(\Gamma)}$	H^1-norm on Γ defined by $\|u\|^2_{H^1(\Gamma)} := \sum_{e\in\mathcal{E}} \|u_e\|^2_{H^1(e)}$
$H^1(\Gamma)$	Sobolev space of square integrable functions with square integrable first derivative on Γ
I	time interval $[0, T]$ with $T \in \mathbb{R}^+$
$L^2([0, T]; H)$	Bochner Space of functions $\boldsymbol{u} : I \to H$
$(\boldsymbol{u}, \boldsymbol{w})_{L^2([0,T];H)}$ ...	inner product on $L^2([0, T]; H)$
$\|\boldsymbol{u}\|_{L^2([0,T];H)}$	norm induced by $(\boldsymbol{u}, \boldsymbol{u})_{L^2([0,T];H)}$

Operators on Graphs

$\mathcal{H}$	standard differential operator $\mathcal{H} : u \mapsto -\frac{\mathrm{d}^2 u}{\mathrm{d}x^2}$
$\mathrm{dom}_{\mathcal{H},\mathrm{NK}}$	domain of $\mathcal{H}$ under Neumann-Kirchhoff boundary conditions
$(\mathcal{K}u)(v) = 0$	abbreviation for the conservation of currents condition on v

Differential Equations on Metric Graphs

f .... right-hand side, usually $f \in L^2(\Gamma)$
ρ .... (positive) potential $\rho \in \mathbb{R}_+$
$\mathcal{R}$ .... reaction term acting on functions u defined on Γ
u^0 .... initial condition of u on Γ
g .... testfunction in $H^1(\Gamma)$
$\mathfrak{h}(u, g)$ .... bilinear form on $H^1(\Gamma)$ defined by $\mathfrak{h}(u, g) := \int_\Gamma \frac{\mathrm{d}u}{\mathrm{d}x}\frac{\mathrm{d}g}{\mathrm{d}x}\mathrm{d}x$
$\mathfrak{h}_\rho(u, g)$ .... bilinear form on $H^1(\Gamma)$ defined by $\mathfrak{h}_\rho(u, g) := \int_\Gamma \frac{\mathrm{d}u}{\mathrm{d}x}\frac{\mathrm{d}g}{\mathrm{d}x}\mathrm{d}x + \rho(u, g)_\Gamma$
$\boldsymbol{u}[t](x)$ .... mapping $\boldsymbol{u} : [0, T] \to H^1(\Gamma)$ associated with $u = u(x, t)$
$|||u||| = |u|_{H^1(\Gamma)}$ ... seminorm induced by the bilinear form $\mathfrak{h}$

Example Graphs

$\Gamma_{\text{star}}, \Gamma_{\text{dia}}$ .... star and diamond graph from Example 2.23
Γ_{BA} .... Barabási-Albert graph from Example 2.24
d .... parameter in the construction of Γ_{BA}
Γ_{ER} .... Erdős-Rényi graph from Example 2.24
p .... parameter in the construction of Γ_{ER}
$\Gamma_{\text{graphene}}, \Gamma_{\text{tree}}$ ...... graphene and tree graph from Example 2.25

Finite Element Method

Discretization and Extended Graphs

h_e .... step size on edge e
$\boldsymbol{h}$ .... collection of step sizes on the edges: $\{h_e\}_{e\in\mathcal{E}}$
$N_e + 1$ .... number of discretization points on edge e
$x_{e,k}$ .... k-th discretization point on edge e, for $k = 1, \ldots, N_e - 1$
$x_{e,0}, x_{e,N_e}$ .... vertex grid points: $x_{e,0} = 0$, $x_{e,N_e} = \ell_e$
$\tilde{\Gamma}_{\boldsymbol{h}}, \tilde{\Gamma}$ .... extended graph of Γ arising from a discretization with step size $\boldsymbol{h}$
$v_{e,k}$ .... k-th discretization point on edge e, interpreted as vertex
$\tilde{\mathcal{V}}, \tilde{\mathcal{V}}_{\boldsymbol{h}}$ .... vertex set of the extended graph
$\tilde{\mathcal{E}}_e$ .... set of partitioned edges belonging to original edge e
$\tilde{\mathcal{E}}$ .... edge set of the extended graph
$\tilde{\mathcal{G}}$ .... underlying combinatorial graph of the extended graph
$\tilde{\mathbf{A}}$ .... adjacency matrix of the extended graph
$\tilde{\mathbf{D}}$ .... degree matrix of the extended graph
$\tilde{\mathbf{L}}$ .... graph Laplacian matrix of the extended graph
$\tilde{\mathbf{N}}$ .... incidence matrix of the extended graph
$\tilde{\mathbf{L}}_{\mathcal{V}\mathcal{V}}, \tilde{\mathbf{L}}_{\mathcal{E}\mathcal{E}}, \tilde{\mathbf{L}}_{\mathcal{E}\mathcal{V}}$ ... blocks of $\tilde{\mathbf{L}}$
$\tilde{\mathbf{L}}_e$ .... block of $\tilde{\mathbf{L}}_{\mathcal{E}\mathcal{E}}$ associated to edge e
$w_{\tilde{e}}$ .... weight assigned to edge $\tilde{e}$, usually $w_{\tilde{e}} = \frac{1}{\ell_{\tilde{e}}}$
$\tilde{\mathbf{W}}$ .... edge-weight matrix of $\tilde{\Gamma}$ defined by $\tilde{\mathbf{W}} = \mathrm{diag}(\{w_{\tilde{e}}\}_{\tilde{e}\in\tilde{\mathcal{E}}})$

$\hat{\mathbf{A}}, \hat{\mathbf{D}}$	weighted adjacency and degree matrix of the extended graph
$\hat{\mathbf{L}}$	weighted graph Laplacian matrix of the extended graph
$\hat{\mathbf{L}}_{\mathcal{V}\mathcal{V}}, \hat{\mathbf{L}}_{\mathcal{E}\mathcal{E}}, \hat{\mathbf{L}}_{\mathcal{V}\mathcal{E}}$	blocks of $\hat{\mathbf{L}}$
$\tilde{\mathbf{N}}_{\mathcal{V}}, \tilde{\mathbf{N}}_{\mathcal{E}}$	blocks of $\tilde{\mathbf{N}}$
$\lvert\mathbf{N}\rvert$	entrywise absolute values of $\mathbf{N}$
$\tilde{\mathbf{N}}^{\text{in}}, \tilde{\mathbf{N}}^{\text{out}}$	incidence matrices of the incoming respectively outgoing edge
$\mathbf{e}_1^{N_e}, \mathbf{e}_{N_e}^{N_e}$	first and last column of the identity matrix of size N_e
$\mathbf{N}_e$	column of $\mathbf{N}$ corresponding to edge e
$N_{\mathcal{E}}$	$N_{\mathcal{E}} = (N_{e_1}, \ldots, N_{e_m})^T$, vector with number of inner grid points per edge
$\hat{\mathbf{S}}$	Schur complement of $\hat{\mathbf{L}}$ in $\hat{\mathbf{L}}_{\mathcal{E}\mathcal{E}}$

Finite Element Approximation

$\psi_{e,k}$	hat basis function on inner discretization point $x_{e,k}$
V_{h_e}	finite element discretization space for edge e
$\mathscr{W}_v$	neighborhood of vertex v
ψ_v	hat basis function defined on $\mathscr{W}_v$
$V_{\boldsymbol{h}}(\Gamma)$	finite element discretization space $V_{\boldsymbol{h}}(\Gamma) \subset H^1(\Gamma)$
$\boldsymbol{u}_{\boldsymbol{h}}(x)$	finite element approximation of $\boldsymbol{u}$
$\mathbf{u}$	vector collecting the coefficients of the finite element discretization
$\mathbf{u}^0$	initial condition in the finite element semidiscretization
$\mathbf{u}_{\mathcal{V}}, \mathbf{u}_{\mathcal{E}}$	blocks of $\mathbf{u}(t) = \left[\mathbf{u}_{\mathcal{V}}(t), \mathbf{u}_{\mathcal{E}}(t)\right]^T$
$\hat{\mathbf{f}}$	finite element discretization of the right hand side f
$\hat{\mathbf{r}}$	finite element discretization of the reaction term $\mathcal{R}$
$\tilde{H}^2(\Gamma)$	direct sum of Sobolev spaces $H^2(e)$ on each edge
$\mathfrak{h}_1(u, g)$	bilinear form on $H^1(\Gamma)$ defined by $\mathfrak{h}_1(u, g) := \int_\Gamma \frac{\mathrm{d}u}{\mathrm{d}x}\frac{\mathrm{d}g}{\mathrm{d}x}\mathrm{d}x + (u, g)_\Gamma$
$\hat{h}$	maximum of step size on all edges, $\hat{h} := \max_{e\in\mathcal{E}} h_e$
$u_{\boldsymbol{h}}, u_{\boldsymbol{h}}^I$	finite element approximation of u, interpolant of u
$c(\Gamma), \hat{c}(\Gamma)$	constants depending on Γ

Multigrid

Δt	time step
$\mathbf{u}^t$	solution of the finite element semidiscretization at time t
J	discretization level
J_0	coarsest discretization level
$\hat{\mathbf{L}}_J, \hat{\mathbf{M}}_J$	finite element discretization matrices on level J
$\mathbf{u}_J^t$	solution of the finite element discretization on level J and time t
$\mathbf{B}, \mathbf{C}, \mathbf{b}$	coefficient matrices and right-hand side of the IMEX discretization

θ	number of time steps computed by a parabolic multigrid iteration
ν	number of smoothing iterations
$\mathcal{S}^\nu$	ν applications of the smoother
η	number of multigrid iterations ($\eta = 1$: V-Cycle, $\eta = 2$: W-Cycle)
$\mathbf{P}, \mathbf{R}$	prolongation and restriction operators

Spectral Solution Method

Trial Functions

$\sigma(\Gamma)$	spectrum of $\mathcal{H}$ acting on Γ with Neumann-Kirchhoff conditions
λ	eigenvalue $\lambda \in \sigma(\Gamma)$
ϕ, ϕ_λ	eigenfunction with $\mathcal{H}\phi = \lambda\phi$, eigenfunction to eigenvalue λ
c_λ	expansion coefficient of u corresponding to ϕ_λ
φ	orthonormal basis of eigenfunctions $\varphi = \{\phi_\lambda : \lambda \in \sigma(\Gamma)\}$
λ_q, ϕ_q, c_q	q-th eigenvalue, eigenfunction and expansion coefficient, enumerated in ascending order $q = 1, 2, 3, \ldots$
X	space spanned by φ
φ_Q	finite basis of eigenfunctions, $\varphi_Q := \{\phi_q, q = 1, \ldots, Q\}$ with $Q \in \mathbb{N}$
X_Q	space spanned by φ_Q
P_Q	L^2-projection on X_Q
$\bar{C}$	positive constant in upper bound on projection coefficients

Spectral Galerkin Approximation

$\boldsymbol{u}_Q$	spectral Galerkin approximation of u, $\boldsymbol{u}_Q \in X_Q$
g_Q	testfunction in X_Q
$\mathbf{c}(t)$	vector collecting the expansion coefficients $c_1(t), \ldots, c_Q(t)$ of $\boldsymbol{u}_Q(t)$
$\mathbf{c}^0$	expansion coefficients of the initial condition, equivalent to $\mathbf{c}(0)$
$\mathbf{\Lambda}$	diagonal matrix of eigenvalues $\lambda_1, \ldots, \lambda_Q$
$\mathbf{f}$	right-hand side of the spectral Galerkin discretization (generalized heat equation)
$\mathbf{r}$	right-hand side of the spectral Galerkin discretization (reaction-diffusion equation)
$\bar{C}_L$	Lipschitz constant of reaction term

Aspects of Implementation

$2h_\kappa$	length of a panel in the Filon quadrature
e_κ	panel on edge e, $e_\kappa := [(\kappa-1)h_\kappa, (\kappa+1)h_\kappa]$
$\tilde{f}_\kappa$	polynomial approximation of f on panel e_κ
ψ	polynomial of degree 2
$\mathcal{I}_e^{\cos}(\lambda_q), \mathcal{I}_e^{\sin}(\lambda_q)$	integrals on edge e corresponding to the sine and cosine part of the eigenfunction ϕ_q
$\boldsymbol{m}_e^{\cos}, \boldsymbol{m}_e^{\sin}$	vector containing the moment integrals of the polynomial approximation corresponding to the sine and cosine part of the eigenfunction ϕ_q on edge e
$\mathcal{F}_e$	vector containing the function evaluations of f on the nodes of the interpolation polynomial on edge e
$\boldsymbol{r}_e$	vector containing the evaluations of the reaction term $\mathcal{R}$ on the nodes of the interpolation polynomial on edge e
$\boldsymbol{\phi}_\kappa^{\cos}, \boldsymbol{\phi}_\kappa^{\sin}$	vectors containing the function evaluations $\cos\left(\sqrt{\lambda_q}\, x\right)$ respectively $\sin\left(\sqrt{\lambda_q}\, x\right)$ on the interpolation node κh_κ
$\mathbf{A}_e, \mathbf{B}_e$	diagonal matrices containing the constants A_e^q, B_e^q of the eigenfunction representation $\phi_q(x) = A_e^q \cos\left(\sqrt{\lambda_q} x\right) + B_e^q \sin\left(\sqrt{\lambda_q} x\right)$

Computation of Quantum Graph Spectra

Relation to Combinatorial Graph Spectra

$\sigma_{\mathcal{V}}(\Gamma)$	vertex spectrum of Γ
$\sigma_{\mathcal{E}}(\Gamma)$	non-vertex spectrum of Γ
$\mathbf{H}$	coefficient matrix of the nonlinear eigenvalue problem associated with $\sigma(\Gamma)$
ϕ_e	eigenfunction on edge e given by $A_e \cos\left(\sqrt{\lambda} x\right) + B_e \sin\left(\sqrt{\lambda} x\right)$
A_e, B_e	edge specific constants of eigenfunction ϕ_e
$\phi_{\mathcal{V}}$	restriction of ϕ to the vertices

Spectra of Equilateral Graphs

$\lambda_{\mu,k}$	Quantum graph eigenvalue associated to discrete eigenvalue μ and $k \in \mathbb{N}$
$\sigma_{\mathcal{V},k}(\Gamma)$	k-th bunch of vertex eigenvalues of Γ
$\tilde{\Delta}_k$	harmonic graph Laplacian matrix of the extended graph with k artificial vertices
$\tilde{\Phi}, \tilde{\mu}$	eigenvector and eigenvalue of $\tilde{\Delta}_k$
$\tilde{\phi}$	eigenfunction of the extended graph
$\phi^{\mathcal{V}}$	vertex eigenfunction
$\phi^{\mathcal{E},\text{odd}}, \phi^{\mathcal{E},\text{even}}$	non-vertex eigenfunctions for k odd and even respectively

Spectra of Non-Equilateral Graphs

$\mathrm{clean}(\Gamma)$	cleaned graph of Γ (all vertices of degree 2 are eliminated)
$\mathfrak{G}_h$	equilateral approximation of Γ with edge length h
$\mathfrak{l}$	length vector of the cleaned graph $\mathrm{clean}(\mathfrak{G})$
$\mathrm{dist}(\Gamma, \mathfrak{G})$	distance of Γ and $\mathfrak{G}$ defined by $\|\boldsymbol{\ell} - \mathfrak{l}\|$
$\mathfrak{G}_h$	equilateral approximation of Γ with edge length h
$\mathfrak{G}_{\mathrm{cl}}, \mathfrak{G}_{\mathrm{fl}}$	equilateral ceil and floor approximation
$\mathfrak{L}_h$	normalized graph Laplacian matrix of $\mathfrak{G}_h$

Chapter 1
Introduction

Abstract This introductory chapter provides a general first overview of metric graphs and partial differential equations on such domains. This is followed by a comprehensive literature review of metric and quantum graphs and an explanation of the scope, novelties, and objectives of this book.

1.1 Motivation

The numerical simulation of phenomena in medicine and biology is a challenging yet aspirational and fascinating task. For this purpose, graphs are an elegant model to represent complex interconnected structures while partial differential equations allow to simulate flows and motion on such a network. In particular, the questioning leading to this work arose from the interdisciplinary research project *"Neurodegeneration Forecasting - A Computational Brainsphere Model for Simulation of Alzheimer's Disease"*.[1] Here, among others, a method to simulate the propagation of misfolded *tau proteins* in the human brain is required. Our approach is to understand the brain as a network of brain regions with certain connectivity patterns. Through these connections, it is assumed that misfolded proteins can spread from one brain region to another, infecting neighboring regions in a *prion* (protein-infection) like fashion.

This assumption is commonly known as the transneuronal spread hypothesis and motivates the modeling of tau propagation as a diffusion process across the brain network. This approach has been prominently addressed by Raj et al. in "A network diffusion model of disease progression in dementia" [45]. There, the brain is considered as a combinatorial graph leading to a discrete model. However, more sophisticated models are required to capture the observed phenomena. For instance, besides the propagation of tau proteins, it is also desirable to consider its aggregation and production via reaction terms. Moreover, since we are considering networks

---

[1] This project was supported by the Excellence Initiative of the University of Cologne from November 2017 to October 2019.

A. Weller, *Numerical Methods for Metric Graphs*, Lecture Notes in Mathematics 2382, https://doi.org/10.1007/978-3-032-05011-3_1

embedded in a three dimensional space (the brain), we wish to interpret them as metric spaces in which each connection has a natural length determined by the distance between its two incident brain regions.

Such topological constructs are known as *metric graphs* in mathematical science. Posing a differential operator along with vertex coupling conditions on a metric graph leads to the prominent concept of *quantum graphs*. There exists a wide range of theoretical literature on quantum graphs, covering topics like spectral theory, semigroup methods and inverse problems. In this work, I will be particularly interested in time dependent problems and therefore prefer to speak of (partial) differential equations (posed) on metric graphs. Due to its prominence, I will nevertheless use the term quantum graph in the context of eigenvalue problems.

More precisely, the main emphasis of this work is on the numerical solution of parabolic equations of the type

$$\frac{\partial}{\partial t}u_e(x,t) - \frac{\partial^2}{\partial x^2}u_e(x,t) = f_e(x,t), \tag{1.1}$$

or, more general, semilinear reaction-diffusion equations

$$\frac{\partial}{\partial t}u_e(x,t) - \frac{\partial^2}{\partial x^2}u_e(x,t) = \mathcal{R}(u_e(x,t)) \tag{1.2}$$

posed on the edges e of a metric graph Γ. In this expression, u_e and f_e are functions defined on the edge e, and $\mathcal{R}(u_e)$ is a (nonlinear) reaction term. The partial differential equations (PDEs) posed on the single edges are coupled at their common vertices, typically by so-called *Neumann-Kirchhoff* conditions

$$\begin{aligned} &u \text{ is continuous on (the vertices of) } \Gamma \\ &\sum_{e\in\mathcal{E}_v} \frac{du_e}{dx}(v) = 0 \quad \text{for all vertices } v \end{aligned} \tag{1.3}$$

where $\mathcal{E}_v$ consists of all edges e incident to vertex v. These "standard vertex conditions" are suitable for our application since they model a continuity and current conservation condition at the vertices of the graph. Theoretical works by von Below [53] and Mugnolo [41] are concerned with such types of (semi-)linear parabolic equations. In particular, existence and uniqueness results have been studied therein.

Besides the already motivated application to brain network models, quantum graphs, or PDEs on metric graphs, are popular models in physical science. The first prominent appearance was a quantum graph model of free electrons in organic molecules in the 1930s [43]. In general, quantum graphs are employed in models of thin branching structures. Nowadays, important applications are, for example, carbon nanostructures in nanotechnogology. For an extensive theoretical investigation and further modeling examples and applications, see for example [6].

1.2 Literature Review

Despite the wide popularity of quantum graphs in modeling as well as in theoretical investigations, the numerical research on PDEs on metric graphs is still in its early stage. From the beginning, this book was inspired by the article "A finite element method for quantum graphs" from Arioli and Benzi [2]. The underlying idea is straightforward: on each edge of the metric graph, the spatial derivative is discretized by linear finite elements, so-called *hat functions*. Some caution is required at the vertices of the graph since the coupling conditions must be satisfied by the solution of the differential equation. In [2], special focus lies on elliptic equations of the type $-\frac{\mathrm{d}^2}{\mathrm{d}x^2}u + \rho u = f$ with a positive potential ρ. A weak formulation is derived and the existence of a solution as well as an error estimate is given. Moreover, the method is applied to a heat equation and a generalized eigenvalue problem on Γ.

Another important aspect of the work of Arioli and Benzi is the representation of the finite element discretization matrices in terms of graph matrices of the so-called *extended graph*. The latter arises by interpreting the spatial grid points as additional vertices in the metric graph. The advantage of this approach is the possibility to assemble the discretization matrix by some manipulations of the incidence matrix of the original graph. Moreover, the structured representation of the discretization matrices enables the application of a domain decomposition approach for the solution of the arising linear systems. For the special case of equilateral graphs (all edges have the same length), an important identity of the Schur complement of the finite element stiffness matrix and the graph Laplacian matrix of the underlying combinatorial graph has been developed.

Recently, the finite element approach was adapted in a work of Stoll and Winkler for optimal control problems on quantum graphs [49].

Besides the finite element approach [2], finite difference approaches have been investigated very recently by Brio et al. [10] and by Besse et al. [7]. In the former, different differential equations involving the second order derivative such as a Poisson's and a wave equation are considered. The metric graph is discretized by placing grid points on both the vertices and the interior of the edges. For the latter inner discretization points, a second order finite difference approximation of the second order derivative is straightforward. A *ghost point* is introduced and subsequently eliminated by means of the Neumann-Kirchhoff conditions to apply the same to the grid points at the vertices. This also ensures the coupling of various edges at their common vertex.

Besse et al. in [7] only place discretization points along the edges and not at the vertices of the graph itself. The vertex conditions are applied to couple the first, respectively, last discretization points of the edges among each other. The approach by Besse et al. [7] was proposed for nonlinear Schrödinger equations on graphs and is implemented in the Python `GraFiDi` Library provided by the same authors. For each vertex, this technique requires the inversion of a matrix of size $\deg \times \deg$ where $\deg$ denotes the number of incident edges of a vertex.

Very recently, there has been further introduced QGLAB, a matlab library for computations on quantum graphs [25]. Here, finite-difference and Chebyshev discretizations are utilized to solve eigenproblems, Poisson problems and linear and nonlinear evolution problems. It further includes the symbolical computation of the graph's secular determinant to derive eigenvalues.

Brio et al. in the same work [10] further suggest a spectral method for the solution of a linear wave and Poisson's equation. This is the attempt most closely related to the one presented as a main part of this book. The basis functions of the spectral expansion are obtained from an eigenvalue problem on the metric graph. An algorithm is proposed for the computation of this spectral basis and proceeds as follows: First, a λ-dependent matrix that is singular at the eigenvalues λ is assembled by coupling the solution of the eigenvalue problem on each edge with Neumann-Kirchhoff conditions. This matrix is constructed by symbolic manipulations and, subsequently, the reciprocal condition number as a function of λ is plotted in order to graphically estimate a lower bound for the distance between two subsequent eigenvalues. The eigenvalues are then determined by a line minimization algorithm applied to subintervals according to the observed distance. The eigenfunctions can be found by computing the null space of the λ-dependent matrix at the associated eigenvalue.

Turning back to the solution of the PDE in the first place, the solution is expanded in terms of the spectral basis and projected onto the eigenfunctions to obtain the unknown coefficients. Spectral decay of the projection coefficients is shown for functions with compact support on each edge. In this case, however, the coupled system reduces to independent equations on the edges since this is equivalent to posing zero Dirichlet conditions at each edge.

1.3 Scope of This Book

The objective of this book is the discussion of numerical methods for the solution of (semi-)linear parabolic PDEs of the type (1.1) or (1.2) posed on metric graphs under Neumann-Kirchhoff conditions (1.3). A suitable framework will be established by introducing initial boundary value problems (IBVPs) governed by the generalized heat equation (1.1) and a semilinear reaction-diffusion equation (1.2), by analyzing the underlying differential operator, and by providing a weak formulation. For the numerical solution, we first discretize in space (*semidiscretization*) and subsequently solve the resulting system of ordinary differential equations (ODEs) by a time stepping method (*method of lines*). For the spatial discretization, we will encounter a Galerkin discretization with finite element trial functions as well as a spectral Galerkin discretization. The semidiscretized systems are then solved by implicit-explicit time stepping methods in the finite element setting and by exponential integrators in the spectral setting.

The application of the finite element method introduced in [2] to the PDEs in focus is analyzed. Its convergence rates are demonstrated by means of simple elliptic and linear parabolic test problems. For the solution of the semidiscretized systems, I propose implicit-explicit time stepping methods which require an efficient solution of the evolving systems of linear equations in each time step. A multigrid ansatz closely related to standard multigrid methods but with suitable graph specific prolongation and restriction operators is presented. As an alternative to the multigrid solver, a generalization of the domain decomposition approach from [2] is derived in the appendix. In particular, the Schur complement identity is also proven for non-equilateral graphs which guarantees a broad application of the method.

The main novelty and focus of this book is the development of a highly accurate spectral Galerkin approximation for the solution of IBVPs governed by (1.1) or (1.2) posed on a metric graph under Neumann-Kirchhoff conditions (1.3) together with the efficient computation of spectral basis functions. The particular characteristic of a spectral solution method is the choice of trial functions in the Galerkin semidiscretization. In contrast to the finite element semidiscretization, where the basis functions typically have local support, orthogonal functions with global support are applied. In the best case, a careful choice of these basis functions guarantees spectral convergence, i.e., faster than any polynomial approximation.

In the studied approach, the trial functions are linear combinations of eigenfunctions obtained from an eigenvalue equation on the metric graph subject to Neumann-Kirchhoff vertex conditions. The ansatz is motivated by the fact that, given the studied coupling conditions, the negative second order derivative is self-adjoint. Therefore, the considered eigenfunctions constitute an orthogonal basis of the solution space. This is equivalent to choosing the eigensolutions of a suitable Sturm-Liouville-Problem (see for example [12]) in the one dimensional setting. We discuss aspects of approximation theory and formulate a spectral Galerkin semidiscretization. Moreover, we address the solution of the semidiscretized system by exponential integrators which necessitates the efficient computation of the appearing function evaluations and integrals. The efficient computation of the latter is also essential to compute the projection coefficients of the spectral expansion and is thereby an important aspect of the proposed approach.

Towards the development of the spectral Galerkin method, it is essential to set up a method to compute eigenvalues and eigenfunctions of quantum graphs. Indeed, the study of eigenvalues and eigenfunctions itself is a very complex and interesting task for which an innovative numerical method is proposed.

I will in this context also speak of the computation of quantum graph spectra in order to keep with the terminology widely used in literature. By this I always mean not only to compute the eigenvalues but also the associated eigenfunctions.

The main result leading to the proposed algorithm is the impressive fact that the continuous eigenvalue equation on a metric graph can be reduced to a *Nonlinear Eigenvalue Problem* (NEP) on the underlying combinatorial graph. This observation has been prominently studied by several authors (for example in [6, 13, 53]) in the special case of equilateral graphs. The problem then actually even simplifies to a linear eigenvalue problem of a graph Laplacian matrix. However, one has to be careful since the relation only applies to a specific part of the spectrum which we refer to as *vertex spectrum*. For the remaining *non-vertex* eigenfunctions, we propose a practical approach by inserting artificial vertices on the edges of the graph.

That said, the spectral method clearly bears some major advantages in the special case that all edges of the metric graph have the same length. In fact, equilateral graphs are important in a wide range of modeling, such as graphene structures. For the general case of non-equilateral graphs, we extend our approach by proposing a numerical method to compute eigenvalues as solutions of the NEP mentioned above. The derived method uses a Newton-trace iteration with the particularity that initial guesses are obtained from suitable equilateral approximations of the non-equilateral graph.

Accompanying this book, I developed the package `MeGraPDE` in `Julia`, available at https://github.com/AnnaWeller/MeGraPDE.jl [54]. Using the methods implemented in `MeGraPDE`, several numerical experiments are presented not only to verify the obtained theoretical results but also to study properties of the solutions and the underlying metric graphs. The finite element and spectral Galerkin discretization are compared in terms of complexity and accuracy by means of a small artificial test problem. Moreover, the applicability of the derived numerical methods to real world, large scale networks is investigated by means of data collected during the interdisciplinary research project on the simulation of Alzheimer's disease. The experiments serve as a starting point for future simulations of tau propagation.

1.4 Outline

Chapter 2 In Sect. 2.1, we give a review of basic concepts of combinatorial graph theory, the representation of graphs through matrices and some relevant results on their spectrum. Given these fundamentals, the concept is extended to define metric graphs in Sect. 2.2. An understanding of function spaces and operators on metric graphs is established, accordingly allowing us to pose parabolic differential equations on metric graphs in Sect. 2.3. We formally define the two central initial boundary value problems associated with a generalized heat and a reaction-diffusion equation. Corresponding weak formulations are derived in Sect. 2.3.2, and their well-posedness is investigated. Finally, in Sect. 2.4, we introduce some concrete graphs and test problems frequently used for illustration purposes and for numerical experiments throughout this work.

Chapter 3 In Chap. 3, the application of the finite element discretization to the IBVPs introduced in Chap. 2 is discussed. We review the discretization of metric graphs and the concept of extended graphs in Sect. 3.1. The finite element discretization of parabolic differential equations in space is outlined in Sect. 3.2 along with the derivation of the semidiscretized system of differential equations and an error analysis. The remainder of the chapter then focusses on the efficient solution of the semidiscretized system with implicit-explicit schemes in combination with a multigrid approach.

Chapter 4 Chapter 4 introduces the heart of this book, the spectral solution method. The trial space for the Galerkin approximation in space is analyzed in Sect. 4.1. We derive an eigenfunction expansion and an estimate for the truncation error. Accordingly, the spectral Galerkin approximation is formalized in Sect. 4.2 yielding a semidiscretized system of ODEs. The remainder of Sect. 4.2 consists of an error analysis before we move on to the solution of the semidiscretized problems in Sect. 4.3. Finally, aspects of implementation, in particular the computation of projections coefficients, are addressed in Sect. 4.3.2.

Chapter 5 Given that the spectral Galerkin discretization in Chap. 4 requires a basis of eigenfunctions, the efficient solution of eigenvalue problems on metric graphs is discussed. The spectrum of a quantum graph is separated in the *vertex* and *non-vertex* spectrum and, for the former, a relation to combinatorial graph spectra is described in Sect. 5.1. In the special case of equilateral graphs, this allows to reduce the continuous eigenvalue equation to a linear eigenvalue problem of a discrete matrix which is used to compute the vertex spectrum in Sect. 5.2.1. In Sect. 5.2.2, we derive a method to compute the remaining part of the spectrum. Some aspects of implementation are discussed in Sect. 5.2.3. In Sect. 5.3, the ideas are adopted to the general class of non-equilateral graphs, and a method to solve the evolving NEP is proposed.

Chapter 6 Chapter 6 is devoted to numerical investigations. The finite element method is addressed in Sect. 6.1. Experiments on the convergence and the solution of the semidiscretized system with the multigrid approach are conducted.

Section 6.2 gives more examples on the computation of quantum graph spectra. On the one hand, we investigate the structure of the spectra. Additionally we test the spectral algorithm in comparison to a finite element approximation of the eigenvalue problem. First large scale examples for the non-equilateral setting are discussed.

The spectral Galerkin method is the subject of Sect. 6.3. Numerical experiments on the projection coefficients, truncation error, and Weyl's law eigenvalue estimates are conducted in Sects. 6.3.1 and 6.3.2 before we move on to investigating spectral solutions of the heat, a fractional diffusion and reaction-diffusion equations in Sects. 6.3.3 and 6.3.5.

This chapter further aims at the comparison of the two previously described methods. We analyze the finite element and spectral Galerkin method with respect to the achieved accuracy and required complexity depending on the degrees of freedom for two small, artificial test problems.

Chapter 7 We give an outlook on the application of the derived methods to the motivational model: the simulation of tau propagation in Alzheimer's disease. The objective of this chapter is to test the applicability of the methods to real world, large scale data. To motivate the application, we introduce the main ideas of the *Global Brainsphere Model* and the representation of the brain as a metric graph. Numerical experiments on the solution of IBVPs governed by linear reaction-diffusion equations are carried out using the finite element discretization from

Chap. 3. The eigenvalue algorithm from Sect. 5.3 is applied to compute eigenvalues of metric graphs modeling the brain.

Chapter 8 I review the main concepts and novelties derived in this manuscript. A brief discussion of the finite difference method, that I briefly mentioned in the literature review, shows how such methods can be embedded in our framework. I further highlight interesting topics for further research that were identified throughout this work.

Chapter 2
Background: Graphs and Differential Equations

Abstract The main concepts of combinatorial, metric and quantum graphs are introduced. Key to this summary is the derivation of a self-adjoint realization of a differential operator on metric graphs. Subsequently, partial differential equations are posed on metric graphs and their weak formulation are derived. In particular, a generalized heat equation and a reaction-diffusion equation on metric graphs are treated. This chapter in addition defines and illustrates several example graphs and test problems frequently used throughout this book.

In order to define continuous partial differential equations on graphs, we need to extend the widely used concept of combinatorial graphs by a metric. Such metric graphs equipped with a differential operator are often called *quantum graphs* and have been studied by several authors in various contexts and under different names since the 1930s, see for example [6, 13, 41, 43, 44], although this list is by far not complete. The survey of combinatorial, metric graphs and quantum graphs is mainly based on the textbook *Introduction to Quantum Graphs* by Berkolaiko and Kuchment [6] which we also recommend for a detailed and comprehensive introduction to a wide range of quantum graph theory.

2.1 Combinatorial Graphs

For the convenience of the reader and to introduce our notation, I give a brief summary on the main concepts of combinatorial graph theory, particularly focusing on graph matrices and their spectral properties required later.

A. Weller, *Numerical Methods for Metric Graphs*, Lecture Notes in Mathematics 2382, https://doi.org/10.1007/978-3-032-05011-3_2

2.1.1 Basic Concepts and Notation

An undirected graph $\mathcal{G} = (\mathcal{V}, \mathcal{E})$ consists of a *vertex* set $\mathcal{V} = \{v_1, \dots, v_n\}$ and a set $\mathcal{E} = \{e_1, \dots, e_m\}$ of unordered pairs of distinct vertices, called the *edges* of $\mathcal{G}$. The number of vertices $|\mathcal{V}|$ is denoted by n and the number of edges $|\mathcal{E}|$ by m. We usually write $e = (v_i, v_j)$ for an edge between the vertices v_i and v_j. Sometimes we use the notations $e_1, \dots, e_m$ to distinguish between several edges or e_{ij} if the endpoints of the edge need to be clarified.

Two distinct vertices v_i and v_j are *adjacent* ($v_i \sim v_j$) if they are connected by an edge $e = (v_i, v_j) \in \mathcal{E}$. Then, we say v_j is a *neighbor* of v_i. The set of all neighbors of v_i is denoted by $\mathcal{N}(v_i)$. We further say the edge $e = (v_i, v_j)$ is *incident* to the vertices v_i and v_j. The number of incident edges of a vertex v is referred to as the *degree* and denoted by $\deg(v)$, the set of incident edges by $\mathcal{E}_v$. We only consider *simple graphs* without loops and multiple edges between two vertices. Thus it holds that $\deg(v) = |\mathcal{N}(v)| = |\mathcal{E}_v|$.

If a vertex has no neighbors in the context of the above definition, it is an *isolated* vertex. Hence, an isolated vertex is not connected to any other vertex in the graph. This leads us to the concept of paths and reachability, which we shortly introduce next.

Consider a sequence of distinct vertices $(v_i, v_{i+1}, \dots, v_j)$ such that all consecutive vertices are connected by an edge, say $e_1 = (v_i, v_{i+1}), e_2 = (v_{i+1}, v_{i+2}), \dots, e_p = (v_{j-1}, v_j)$. The sequence of distinct edges $(e_1, \dots, e_p)$ is a *path* of length p between v_i and v_j. If such a sequence exists, v_i and v_j are connected, or v_j is reachable from v_i. In the case of identical start point v_i and end point v_j, we speak of a *cycle*. The graph is *connected* if there exists a path between every pair of two vertices. In this work, we always assume a graph to be connected. Otherwise, one can treat the connected components as separate graphs and apply the same theory straightaway.

Shortest paths between vertices naturally introduce a kind of distance on graphs, i.e., the distance $\text{dist}(v_i, v_j)$ between two vertices is the length of the shortest path connecting v_i and v_j. Clearly, $\text{dist}(v_i, v_j) = 1$ if and only if $v_i \sim v_j$.

Although we only consider undirected combinatorial graphs, it is convenient for the definition of metric graphs, later on, to introduce orientations on edges. By orientation, we mean that we assign a direction to each edge. Such directed edges are usually called *bonds* in the context of *digraphs* (directed graphs). The start- and endpoint of a bond b are often identified by $o(b)$ and $t(b)$ which stands for *origin* and *terminal* vertex. The reversal of a bond b with opposite direction is denoted by $\bar{b}$ with $o(\bar{b}) = t(b)$ and $t(\bar{b}) = o(b)$.

Throughout this work, we try to omit the term bond as our discussed model graphs are always undirected graphs and the orientation only serves the technical purpose of uniquely defining coordinates along the edges. In a slight abuse of notation, we therefore write $o(e)$ and $t(e)$ for the start- and endpoint of an oriented edge. In this context, we use the notation $\mathcal{E}_v^{\text{out}} := \{e \in \mathcal{E} : v = o(e)\}$ for the set of outgoing edges of v and $\mathcal{E}_v^{\text{in}} := \{e \in \mathcal{E} : v = t(e)\}$ for the set of incoming edges.

2.1.2 Graph Matrices and Eigenvalues

We introduce the characterization of graphs through graph matrices, which are essential for our further work.

The *adjacency matrix* $\mathbf{A} \in \mathbb{R}^{n \times n}$ is defined by $\mathbf{A} := (a_{ij})_{i,j=1,\dots,n}$ with $a_{ij} = 1$ if $v_i \sim v_j$ and 0 otherwise. If a graph is simple, it can be completely characterized by its adjacency matrix. We further define the *degree matrix* $\mathbf{D} := \mathrm{diag}(\deg(v_1), \dots, \deg(v_n))$ and the *graph Laplacian matrix* $\mathbf{L} := \mathbf{D} - \mathbf{A}$.

A graph can also be characterized by its *incidence matrix* $\mathbf{N} \in \mathbb{R}^{n \times m}$. Here, the rows of the matrix correspond to the vertices $v_1, \dots, v_n$, and the columns to the edges $e_1, \dots, e_m$ with the entries describing their incidence. After assigning an arbitrary but fixed orientation to the edges, we can define

$$(\mathbf{N})_{v,e} := \begin{cases} 1, & \text{if } v = t(e) \\ -1, & \text{if } v = o(e) \\ 0, & \text{otherwise} \end{cases}$$

and obtain the useful relation $\mathbf{L} = \mathbf{N}\mathbf{N}^T$. In the further course of the work, we use the convention that the first entry per row of the incidence matrix is always negative. This is equivalent to orientating the edges from the smaller to the larger vertex index.

With this notation in place, we prove the following well-known theorem on the spectrum of $\mathbf{L}$, see for example [11].

Theorem 2.1 *Let $\mathcal{G} = (\mathcal{V}, \mathcal{E})$ be a simple connected graph with graph Laplacian matrix $\mathbf{L}$. Then, $\mathbf{L}$ is positive semidefinite with a simple zero eigenvalue corresponding to the eigenvector $\mathbf{1} := (1, \dots, 1)^T$.*

Proof For $\mathbf{u} \in \mathbb{R}^n$, let $\mathbf{u}(v_i)$ denote the i-th entry[1] of $\mathbf{u}$ and consider

$$\mathbf{u}^T \mathbf{L} \mathbf{u} = \sum_{(v_i, v_j) \in \mathcal{E}} (\mathbf{u}(v_i) - \mathbf{u}(v_j))^2 \geq 0$$

which implies that $\mathbf{L}$ is positive semidefinite. This is also obvious from the fact that $\mathbf{L}$ is weakly diagonally dominant. It remains to show that the null space is spanned by the eigenvector $\mathbf{1} := (1, \dots, 1)^T$. Obviously, $\mathbf{L}\mathbf{1} = 0$ and, on the other hand, for $\mathbf{u} \in \mathbb{R}^n$ with

$$\mathbf{u}^T \mathbf{L} \mathbf{u} = \sum_{(v_i, v_j) \in \mathcal{E}} (\mathbf{u}(v_i) - \mathbf{u}(v_j))^2 = 0,$$

[1] This notation refers to the interpretation of $\mathbf{u}$ as a function on the vertices of $\mathcal{G}$, as introduced later.

we obtain $\mathbf{u}(v_i) = \mathbf{u}(v_j)$ for all $v_i \sim v_j$. Since $\mathcal{G}$ is connected, we conclude that $\mathbf{u}$ must be constant across all vertices. □

Of special interest for the analysis of the spectrum of quantum graphs are the *harmonic graph Laplacian matrix* $\Delta_{\mathcal{G}}$ and the *normalized graph Laplacian matrix* $\mathcal{L}$ defined by

$$\Delta_{\mathcal{G}} := \mathbf{D}^{-1}\mathbf{L} \quad \text{and} \quad \mathcal{L} := \mathbf{D}^{-\frac{1}{2}}\mathbf{L}\mathbf{D}^{-\frac{1}{2}}.$$

As in the notation $\Delta_{\mathcal{G}}$, we will sometimes work with a subscript if the underlying graph is not clear from the context, for example, $\mathcal{L}_{\mathcal{G}_1}$ and $\mathcal{L}_{\mathcal{G}_2}$ are the normalized graph Laplacian matrices of $\mathcal{G}_1$ and $\mathcal{G}_2$ respectively.

Since $\Delta_{\mathcal{G}} = \mathbf{D}^{-\frac{1}{2}}\mathcal{L}\mathbf{D}^{\frac{1}{2}} = \mathbf{I} - \mathbf{D}^{-\frac{1}{2}}(\mathbf{I} - \mathcal{L})\mathbf{D}^{\frac{1}{2}}$, we can deduce that $\Delta_{\mathcal{G}}$ and $\mathcal{L}$ are similar and consequently have the same eigenvalues $\mu_i, i = 1, \ldots, n$. Moreover, if Φ is eigenvector of $\Delta_{\mathcal{G}}$, it holds that $\mathbf{D}^{\frac{1}{2}}\Phi$ is eigenvector of $\mathcal{L}$. These observations will be particularly useful to compute the eigenvalue decomposition of $\Delta_{\mathcal{G}}$, as the normalized graph Laplacian matrix is, in contrast to the harmonic graph Laplacian matrix, symmetric. Furthermore, the following well-known estimate on μ_i, which can be found in [15], Lemma 1.7, will play an essential role in the following chapters.

Theorem 2.2 *Let $\mathcal{G} = (\mathcal{V}, \mathcal{E})$ be a simple connected graph with normalized graph Laplacian matrix $\mathcal{L}$ and harmonic graph Laplacian $\Delta_{\mathcal{G}}$. Then, for the eigenvalues μ_i of $\mathcal{L}$ and $\Delta_{\mathcal{G}}$ it holds that*

$$0 \le \mu_i \le 2$$

for all $i = 1, \ldots, n$ with $\mu_n = 2$ if and only if $\mathcal{G}$ is bipartite. Moreover, both $\mu_1 = 0$ and $\mu_n = 2$ (if it occurs) are simple eigenvalues.

Remark 2.3 The spectrum of the normalized and harmonic graph Laplacian matrices will be denoted by $\sigma(\mathcal{L})$ and $\sigma(\Delta_{\mathcal{G}})$ respectively.

Before we prove the theorem, we briefly give the definition of a bipartite graph.

Definition 2.4 A graph $\mathcal{G} = (\mathcal{V}, \mathcal{E})$ is bipartite, if there exists a disjoint partitioning of the vertices $\mathcal{V} = \mathcal{V}_1 \cup \mathcal{V}_2$ such that there are no edges inside a partition, i.e., $v_i \not\sim v_j$ for all vertices $v_i, v_j \in \mathcal{V}_1$ and $v_i \not\sim v_j$ for all vertices $v_i, v_j \in \mathcal{V}_2$.

In particular, a bipartite graph has a two-coloring, that is, the vertices can be colored with two colors such that every two adjacent vertices have a different color, see Fig. 2.1. We may now turn to the proof of Theorem 2.2.

Proof of Theorem 2.2 The proof follows [15] and was prepared in the context of a seminar paper[2] according to our background and notation. Due to the importance

[2] Schmitz L., Stiller, D.: Spectra of quantum graphs. Seminar paper in the summer semester 2022, Mathematical Institute, University of Cologne.

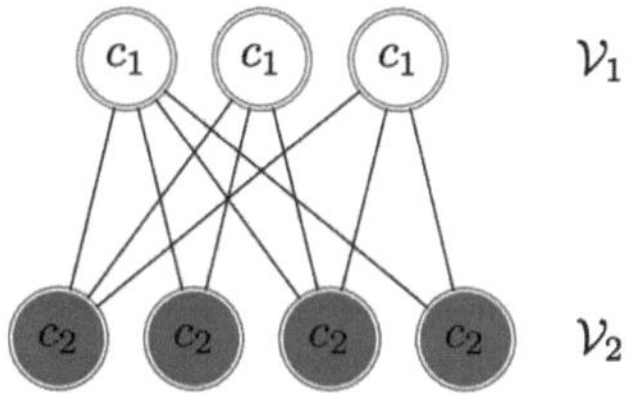

Fig. 2.1 Bipartite graph colored with two colors c_1 and c_2

of Theorem 2.2, we include it here for the sake of completeness. As $\Delta_{\mathcal{G}}$ and $\mathcal{L}$ are similar, it suffices to prove the theorem for $\mathcal{L}$ and remark that $\Delta_{\mathcal{G}}$ has the same eigenvalues.

Let (μ, Υ) be an eigenpair of $\mathcal{L}$. Then,

$$\mu\Upsilon = \mathcal{L}\Upsilon = \mathbf{D}^{-\frac{1}{2}}\mathbf{L}\mathbf{D}^{-\frac{1}{2}}\Upsilon$$

is equivalent to

$$\mu\mathbf{D}^{-\frac{1}{2}}\Upsilon = \Delta_{\mathcal{G}}\mathbf{D}^{-\frac{1}{2}}\Upsilon.$$

Thus, $\mu\Phi = \Delta_{\mathcal{G}}\Phi$ where $\Phi = \mathbf{D}^{-\frac{1}{2}}\Upsilon$ is the eigenvector of the harmonic graph Laplacian corresponding to the eigenvalue μ. Since $\mathcal{L}$ is real and symmetric, we can characterize the eigenvalues through the Rayleigh quotient:

$$\mu = \frac{\Upsilon^T\mathcal{L}\Upsilon}{\Upsilon^T\Upsilon} = \frac{\Upsilon^T\mathbf{D}^{-\frac{1}{2}}\mathbf{L}\mathbf{D}^{-\frac{1}{2}}\Upsilon}{\Upsilon^T\Upsilon} = \frac{\Phi^T\mathbf{L}\Phi}{(\mathbf{D}^{\frac{1}{2}}\Phi)^T(\mathbf{D}^{\frac{1}{2}}\Phi)}$$
$$= \frac{\sum_{(v_i,v_j)\in\mathcal{E}}(\Phi(v_i) - \Phi(v_j))^2}{\sum_{v_i\in\mathcal{V}}\deg(v_i)\Phi(v_i)^2}.$$

Clearly, it follows that $\mu \geq 0$ and $\mu = 0$ if and only if $\Phi(v_i) = \Phi(v_j)$ for all $v_i, v_j \in \mathcal{V}$ equivalent to the proof of Theorem 2.1. For the largest eigenvalue μ_n it holds that

$$\mu_n = \sup \frac{\sum_{(v_i,v_j)\in\mathcal{E}}(\Phi(v_i) - \Phi(v_j))^2}{\sum_{v_i\in\mathcal{V}}\deg(v_i)\Phi(v_i)^2}.$$

For the numerator, we derive

$$\sum_{(v_i,v_j)\in\mathcal{E}} (\Phi(v_i) - \Phi(v_j))^2 \leq 2 \sum_{(v_i,v_j)\in\mathcal{E}} \Phi(v_i)^2 + \Phi(v_j)^2$$
$$= 2\frac{1}{2}\sum_{v_i\in\mathcal{V}}\sum_{v_j\sim v_i} \Phi(v_i)^2 + \Phi(v_j)^2$$

$$= \sum_{v_i \in \mathcal{V}} \deg(v_i)\Phi(v_i)^2 + \sum_{v_i \in \mathcal{V}} \sum_{v_j \sim v_i} \Phi(v_j)^2$$

$$= \sum_{v_i \in \mathcal{V}} \deg(v_i)\Phi(v_i)^2 + \sum_{v_j \in \mathcal{V}} \deg(v_j)\, \Phi(v_j)^2$$

$$= 2 \sum_{v_i \in \mathcal{V}} \deg(v_i)\Phi(v_i)^2.$$

Together, we obtain

$$\mu_n = \sup \frac{\sum_{(v_i,v_j)\in\mathcal{E}} (\Phi(v_i) - \Phi(v_j))^2}{\sum_{v_i\in\mathcal{V}} \deg(v_i)\Phi(v_i)^2} \le 2 \sup \frac{\sum_{v_i\in\mathcal{V}} \deg(v_i)\Phi(v_i)^2}{\sum_{v_i\in\mathcal{V}} \deg(v_i)\Phi(v_i)^2} = 2.$$

Moreover, equality holds if and only if

$$\sum_{(v_i,v_j)\in\mathcal{E}} (\Phi(v_i) - \Phi(v_j))^2 = 2 \sum_{(v_i,v_j)\in\mathcal{E}} \left(\Phi(v_i)^2 + \Phi(v_j)^2\right),$$

which is equivalent to

$$(\Phi(v_i) - \Phi(v_j))^2 = \Phi(v_i)^2 - 2\Phi(v_i)\Phi(v_j) + \Phi(v_j)^2 = 2(\Phi(v_i)^2 + \Phi(v_j)^2)$$

for all $v_i, v_j \in \mathcal{V}$ with $v_i \sim v_j$. This can only be true if we choose $\Phi(v_i) = -\Phi(v_j)$ for all $v_i \sim v_j$, which is possible if and only if $\mathcal{G}$ can be colored with two colors, i.e., if $\mathcal{G}$ is bipartite. □

2.2 Metric Graphs

In the context of combinatorial graphs, the edges can be any kind of (abstract) relation, for example, a friendship in a social network. However, in the scope of this work, we understand an edge e as a physical connection between two vertices with a given length ℓ_e. Therefore, we identify each edge with an interval $[0, \ell_e]$. The intervals are coupled by their start and end vertex just as in the combinatorial graph. In particular, the point $x_e = 0$ is identified with the start vertex $o(e)$ and $x_e = \ell_e$ with the end vertex $t(e)$. We now see what purpose the orientation of the edges serves, namely to uniquely define the coordinates on the edges. The points of a metric graph are then not only the vertices $v \in \mathcal{V}$ but all points $x_e \in [0, \ell_e]$, $e \in \mathcal{E}$ on the edges. We summarize this concept in the following definition.

Definition 2.5 A metric graph Γ is defined by a combinatorial graph $\mathcal{G} = (\mathcal{V}, \mathcal{E})$ and a vector of edge lengths $\boldsymbol{\ell} := (\ell_{e_1}, \ldots, \ell_{e_m})^T \in \mathbb{R}^m$ with $\ell_e \in \mathbb{R}^+$ for all $e \in \mathcal{E}$ by

(a) assigning an (arbitrary but fixed) orientation to each edge e,
(b) assigning the length ℓ_e to each edge e,
(c) parameterizing each edge by $x_e \in [0, \ell_e]$ where the coordinate x_e is increasing in the direction of e.

We write $x \in \Gamma$ if $x \in \sqcup_{e\in\mathcal{E}} [0, \ell_e]$, and we use the subscript x_e to clarify on which edge e the point lies. Identifying e with an interval $e = [0, \ell_e]$, the distance between two vertices $v_i, v_j \in \Gamma$ is defined by $\ell_1 + \ldots + \ell_p$ where $(e_1, \ldots, e_p)$ is the shortest path connecting v_i and v_j. This concept can be extended to $x, y \in \Gamma$ to define the distance between two arbitrary points. We will sometimes need the total length of a metric graph which we refer to as *volume* of Γ, i.e., $\mathrm{vol}_\Gamma := \sum_{e\in\mathcal{E}} \ell_e$.

Remark 2.6 Usually, we assume the orientation of the edges to be assigned according to the index of the vertices, i.e., such that $o(e_{ij}) = v_i$ if $i < j$. However, whenever we consider an arbitrary but fixed vertex v, we sometimes assume v to be the origin of all incident edges. This helps to simplify the notation since v is then identified with $x_e = 0$ for all $e \in \mathcal{E}_v$, see also Fig. 2.2 for an illustration.

Within the framework of this book, we only consider *finite metric graphs* with finite edge lengths $\ell_e < \infty$ and a finite number of edges and vertices. These finite metric graphs are compact if interpreted as a topological space and therefore often referred to as *compact metric graphs*. As a particular class of metric graphs, we introduce *equilateral graphs*.

Definition 2.7 A metric graph Γ is *equilateral* if each edge has the same length ℓ, i.e., $\ell_e = \ell$ for all $e \in \mathcal{E}$.

These graphs play an important role in the determination of spectra of differential operators posed on Γ. Moreover, we will often use equilateral graphs for demonstration purposes since they allow a simplified notation.

We sometimes need to distinguish between metric graphs with underlying bipartite and non-bipartite combinatorial graphs which is why I give the following definition.

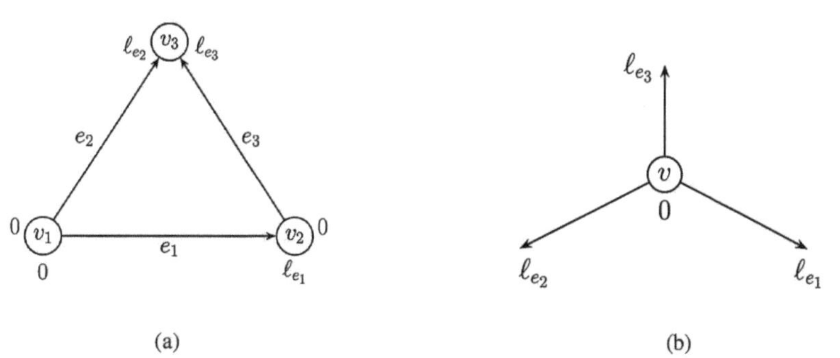

Fig. 2.2 Orientation of edges in a metric graph. (**a**) Classical orientation of edges from smaller to larger vertex index. (**b**) Simplified orientation for fixed vertex v

Definition 2.8 A metric graph Γ is referred to as *bipartite*, if the underlying combinatorial graph $\mathcal{G}$ is bipartite.

2.2.1 Function Spaces on Graphs

A function $\mathbf{u} : \mathcal{V} \to \mathbb{R}$ defined on a combinatorial graph $\mathcal{G}$ assigns a value to each vertex $v \in \mathcal{V}$ and thus can be identified with a vector in $\mathbb{R}^n$. For metric graphs, we need an extension of the notion of functions defined on both the vertices and edges.

Definition 2.9 A function $u : \Gamma \to \mathbb{R}$ on a metric graph is a collection $\{u_e\}_{e \in \mathcal{E}}$ of functions $u_e : [0, \ell_e] \to \mathbb{R}$ defined on the edges $e \in \mathcal{E}$.

We use both the notation $u(x)$ to evaluate u at an arbitrary point $x \in \Gamma$ and $u(x_e)$ for $x_e \in [0, \ell_e]$ if we refer to a point on a specific edge e. Equivalently, we can use the expression $u_e(x)$ for $x \in [0, \ell_e]$ which also clarifies that we are considering a point on a specific edge e.

Before defining function spaces on metric graphs, we first need to clarify what we understand by continuous functions on metric graphs.

Definition 2.10 A function $u : \Gamma \to \mathbb{R}$ is *continuous* on Γ if $u_e : [0, \ell_e] \to \mathbb{R}$ is continuous for all $e \in \mathcal{E}$ and for a given vertex v it holds that $u_e(0)$ assumes the same value for all $e \in \mathcal{E}_v$.

For simplicity of notation, we here assumed all edges incident to v to be oriented in the direction away from v. In other words, the values of the functions defined across the single edges agree at the endpoints. The space of continuous functions on Γ is denoted by $C(\Gamma)$. In particular, it is important to notice that for $u \in C(\Gamma)$, the expression $u(v)$ as well as the restriction of u to the vertices $\mathbf{u}_{\mathcal{V}} := (u(v_1), \dots, u(v_n))^T$ are well defined. We now consider the following function spaces on metric graphs as introduced in [6].

Definition 2.11

(a) The space of square integrable, measurable functions on Γ is given by

$$L^2(\Gamma) := \bigoplus_{e \in \mathcal{E}} L^2(e) \quad \text{with} \quad \|u\|_\Gamma^2 := \sum_{e \in \mathcal{E}} \|u_e\|_{L^2(e)}^2$$

where $\|u_e\|_{L^2(e)}^2 = \langle u_e, u_e \rangle_e$ and $\langle u_e, w_e \rangle_e = \int_0^{\ell_e} u_e(x) w_e(x)\, \mathrm{d}x$ for $u_e, w_e \in L^2(e)$. The corresponding $L^2(\Gamma)$ inner product on Γ is defined by

$$(u, w)_\Gamma := \int_\Gamma u(x) w(x) \mathrm{d}x = \sum_{e \in \mathcal{E}} \int_0^{\ell_e} u_e(x) w_e(x) \mathrm{d}x.$$

(b) The Sobolev space $H^1(\Gamma)$ of square integrable functions with square integrable first derivative on Γ is given by

$$H^1(\Gamma) := \bigoplus_{e\in\mathcal{E}} H^1(e) \cap C(\Gamma) \quad \text{with} \quad \|u\|^2_{H^1(\Gamma)} := \sum_{e\in\mathcal{E}} \|u_e\|^2_{H^1(e)}.$$

Here, the square of the norm on $H^1(e)$ is defined by $\|u_e\|^2_{L^2(e)} + \|u'_e\|^2_{L^2(e)}$ where u'_e denotes the first derivative of u_e on the edge e with respect to the parameter $x \in [0, \ell_e]$.

Of special interest will further be parabolic differential equations on Γ. The definition of *Bochner Spaces* on metric graphs is then standard (see for example [22], Section 5.9.2):

Definition 2.12 Let H be either $L^2(\Gamma)$ or $H^1(\Gamma)$ and $I := [0, T] \subset \mathbb{R}$ an interval. The Bochner Space $L^2([0, T]; H)$ is the space of functions $\boldsymbol{u} : I \to H$ such that the norm

$$\|\boldsymbol{u}\|_{L^2(I;H)} := \left(\int_I \|\boldsymbol{u}(t)\|^2_H \mathrm{d}t\right)^{\frac{1}{2}}$$

is finite. Moreover, we define the space $C([0, T]; H)$ of continuous functions $\boldsymbol{u} : I \to H$ with

$$\|\boldsymbol{u}\|_{C(I;H)} := \max_{0\le t\le T} \|\boldsymbol{u}(t)\|_H < \infty.$$

2.2.2 *Operators on Graphs*

In the setting of combinatorial graphs, all graph matrices introduced in Sect. 2.1.2 can also be understood as discrete operators acting on functions defined on $\mathcal{G}$. For example, the graph Laplacian matrix $\mathbf{L}$ can be associated with the operator

$$(\mathbf{Lu})(v_i) = \sum_{v_j\in\mathcal{V}} \mathbf{L}_{ij}\mathbf{u}(v_j) = \deg(v_i)\mathbf{u}(v_i) - \sum_{v_j\sim v_i} \mathbf{u}(v_j),$$

see [6] for a detailed discussion. In this context, some authors speak of the graph Laplacian matrix to be the discrete analogous (up to the sign) of the standard Laplace operator on a graph and the incidence matrix $\mathbf{N}$ to be the discrete analog of the exterior derivative, which is also reflected in the relationship $\mathbf{L} = \mathbf{N}\mathbf{N}^T$ [6].

Since the main motivation of this work are diffusion-type equations on metric graphs, we are interested in the *standard differential operator* $\mathcal{H}$ acting as

$$\mathcal{H} : u \mapsto -\frac{\mathrm{d}^2 u}{\mathrm{d}x^2}, \tag{2.1}$$

or, in a more formal definition,

$$\mathcal{H} : \{u_e, e \in \mathcal{E}\} \mapsto \left\{ -\frac{\mathrm{d}^2}{\mathrm{d}x^2} u_e,\ e \in \mathcal{E} \right\},$$

i.e., the negative second order derivative acting on each edge. We will consider $\mathcal{H}$ on the domain introduced in the following definition.

Definition 2.13 The *Neumann-Kirchhoff vertex conditions* are defined as the pair of coupling conditions

$$u \text{ is continuous on } \Gamma \tag{2.2a}$$

$$\sum_{e \in \mathcal{E}_v} \frac{\mathrm{d}u_e}{\mathrm{d}x}(v) = 0 \quad \text{for all } v \in \mathcal{V} \tag{2.2b}$$

where the derivatives in (2.2b) are assumed to be taken in the direction away from the vertex. Of particular interest are functions in

$$\mathrm{dom}_{\mathcal{H},\mathrm{NK}} := \bigoplus_{e \in \mathcal{E}} H^2(e) \cap \{u \text{ fulfills Neumann-Kirchhoff conditions}\}. \tag{2.3}$$

The Neumann-Kirchhoff conditions are appropriate for our modeling purpose since (2.2b) physically corresponds to a current conservation condition and, together with the continuity condition (2.2a), these are the natural demands on a concentration function. In order to easily distinguish between the two conditions throughout this work, we introduce the shorthand notation as seen in [49]:

$$(\mathcal{K}u)(v) := \sum_{e \in \mathcal{E}_v} \frac{\mathrm{d}u_e}{\mathrm{d}x}(v). \tag{2.4}$$

Neumann-Kirchhoff conditions are more or less the most common vertex conditions considered in literature and imply some useful properties of $\mathcal{H}$. In particular, the following theorem is a well-known implication, see for example [6], Theorem 1.4.4 and (1.4.26).

Theorem 2.14 *The negative second order derivative* $\mathcal{H} : \mathrm{dom}_{\mathcal{H},\mathrm{NK}} \subset L^2(\Gamma) \to L^2(\Gamma)$ *is self-adjoint.*

We conclude this subsection with the definition of the commonly used concept of *Quantum Graphs*.

Definition 2.15 A *Quantum Graph* is a metric graph Γ equipped with a differential operator $\mathcal{H}$ and coupling conditions at the vertices of Γ.

Note that some authors require self-adjointness of $\mathcal{H}$ in the definition of a quantum graph. Since we will be only interested in the standard differential operator $\mathcal{H} : u \mapsto -\frac{\mathrm{d}^2 u}{\mathrm{d}x^2}$ and Neumann-Kirchhoff conditions, a quantum graph in the scope of this work reduces to the triple

$$\{\text{metric graph } \Gamma, \mathcal{H} : u \mapsto -\frac{\mathrm{d}^2 u}{\mathrm{d}x^2}, \text{ Neumann-Kirchhoff conditions}\}.$$

2.3 Differential Equations on Metric Graphs

Oftentimes, eigenvalue equations on metric graphs are considered. For the standard differential operator

$$\mathcal{H} : u \mapsto -\frac{\mathrm{d}^2 u}{\mathrm{d}x^2}$$

with Neumann-Kirchhoff conditions, the eigenvalue equation

$$\mathcal{H}u = \lambda u \tag{2.5}$$

on the metric graph is nothing else than the system of differential equations

$$-\frac{\mathrm{d}^2}{\mathrm{d}x^2} u_e = \lambda u_e \quad \text{for all } e \in \mathcal{E}$$

with coupling conditions

$$u \text{ is continuous on } \Gamma$$
$$(\mathcal{K}u)(v) = 0 \text{ for all } v \in \mathcal{V}.$$

Similarly, one can pose more general, elliptic differential equations on Γ, for example

$$\mathcal{H}u = f \text{ on } \Gamma$$

or

$$\mathcal{H}u + \rho u = f \text{ on } \Gamma,\ \rho \in \mathbb{R}^+$$

each with Neumann-Kirchhoff conditions and $f \in L^2(\Gamma)$. More generally, ρ can also be a real-valued function $\rho \in L^\infty(\Gamma) := \bigoplus_{e\in\mathcal{E}} L^\infty(e)$.

2.3.1 *Parabolic PDEs on Metric Graphs*

We are particularly interested in diffusion-type equations posed on metric graphs. In this context, we now deal with functions also depending on a time variable t, i.e.,

$$u(x,t) : \Gamma \times [0,T] \to \mathbb{R}$$

where $[0, T]$ is an interval with $T \in \mathbb{R}^+$. Consider for example the linear parabolic equation

$$\frac{\partial u}{\partial t} + \mathcal{H}u = f \quad \text{on } \Gamma \times [0,T]$$

with $f : \Gamma \times [0,T] \to \mathbb{R}$. In the special case $f \equiv 0$ we recover the *heat equation* on a metric graph. Together with vertex coupling and initial conditions, we define the following initial boundary value problem.

Problem 2.16 (IBVP for Generalized Heat Equation) Let Γ be a compact, connected metric graph, $\mathcal{H} : u \mapsto -\frac{\partial^2 u}{\partial x^2}$, $T > 0$ and $f : \Gamma \times [0,T] \to \mathbb{R}$. We study the *initial boundary value problem for the generalized heat equation* on Γ given by

$$\begin{aligned} \frac{\partial u}{\partial t} + \mathcal{H}u &= f && \text{on } \Gamma \times [0,T], \\ u(\cdot,t) \text{ is continuous} & && \text{on } \Gamma \text{ for } t \in [0,T], \\ (\mathcal{K}u(\cdot,t))(v) &= 0 && \text{for all } v \in \mathcal{V},\ t \in [0,T], \\ u(\cdot,0) &= u^0 && \text{on } \Gamma \end{aligned} \tag{2.6}$$

with initial condition $u^0 \in \mathrm{dom}_{\mathcal{H},\mathrm{NK}}$.

More general, we consider reaction-diffusion equations

$$\frac{\partial u}{\partial t} + \mathcal{H}u = \mathcal{R}(u) \quad \text{on } \Gamma \times [0,T]. \tag{2.7}$$

We will always assume that the reaction term $\mathcal{R}$ is Lipschitz continuous. However, if $\mathcal{R}$ is not linear, (2.7) is a semilinear parabolic equation on Γ. In either case, we define the corresponding initial boundary value problem as follows.

Problem 2.17 (IBVP for Reaction-Diffusion Equation) Let Γ be a compact, connected metric graph, $\mathcal{H} : u \mapsto -\frac{\partial^2 u}{\partial x^2}$, $T > 0$ and $\mathcal{R} : \mathbb{R} \to \mathbb{R}$ Lipschitz continuous. We consider the *initial boundary value problem for the reaction-diffusion equation* on Γ given by

$$\begin{aligned} \frac{\partial u}{\partial t} + \mathcal{H}u &= \mathcal{R}(u) && \text{on } \Gamma \times [0, T], \\ u(\cdot, t) \text{ is continuous} & && \text{on } \Gamma \text{ for } t \in [0, T], \\ (\mathcal{K}u(\cdot, t))(v) &= 0 && \text{for all } v \in \mathcal{V},\ t \in [0, T], \\ u(\cdot, 0) &= u^0 && \text{on } \Gamma \end{aligned} \tag{2.8}$$

with initial condition $u^0 \in \mathrm{dom}_{\mathcal{H},\mathrm{NK}}$.

The considered problems fit into the category of semilinear parabolic PDEs for which well-posedness has been studied by von Below in his dissertation [52]. We point out to the interested reader that evolution equations on metric graphs have also been investigated more recently by Mugnolo [41] who uses semigroup methods.

2.3.2 *Weak Formulations*

For $u, g \in H^1(\Gamma)$, we define the bilinear form

$$\mathfrak{h}(u, g) := \int_\Gamma \frac{\mathrm{d}u}{\mathrm{d}x}\frac{\mathrm{d}g}{\mathrm{d}x}\mathrm{d}x. \tag{2.9}$$

Before we establish weak formulations of PDEs on metric graphs, we prove the following integration by parts formula.

Theorem 2.18 *Let $g \in H^1(\Gamma)$, $u \in \bigoplus_{e\in\mathcal{E}} H^2(e)$ and $\mathcal{H}$ be the standard differential operator $\mathcal{H} : u \mapsto -\frac{\mathrm{d}^2 u}{\mathrm{d}x^2}$. Then, we obtain the following integration by parts formula on Γ:*

$$(\mathcal{H}u, g)_\Gamma = \mathfrak{h}(u, g) + \sum_{v\in\mathcal{V}} g(v) \sum_{e\in\mathcal{E}_v} \frac{\mathrm{d}u_e}{\mathrm{d}x}(v). \tag{2.10}$$

Note that whenever we use the expression $\frac{\mathrm{d}u_e}{\mathrm{d}x}(v)$ for a vertex $v \in \mathcal{V}$, we assume the derivative to be taken in the direction away from the vertex v (as in the Neumann-

Kirchhoff conditions). The proof was prepared for an internal manuscript [3] and is included here for the sake of completeness.

Proof By definition, $(\mathcal{H}u, g)_\Gamma = -\int_\Gamma \frac{\mathrm{d}^2 u}{\mathrm{d}x^2}(x)\, g(x)\, \mathrm{d}x = -\sum_{e\in\mathcal{E}} \int_0^{\ell_e} \frac{\mathrm{d}^2 u_e}{\mathrm{d}x^2}(x) g_e(x) \mathrm{d}x$. Integration by parts yields

$$-\sum_{e\in\mathcal{E}} \int_0^{\ell_e} \frac{\mathrm{d}^2 u_e}{\mathrm{d}x^2}(x) g_e(x) \mathrm{d}x = \sum_{e\in\mathcal{E}} \int_0^{\ell_e} \frac{\mathrm{d}u_e}{\mathrm{d}x}(x) \frac{\mathrm{d}g_e}{\mathrm{d}x}(x) \mathrm{d}x - \sum_{e\in\mathcal{E}} \left[\frac{\mathrm{d}u_e}{\mathrm{d}x}(x) g_e(x)\right]_0^{\ell_e}.$$

Again by definition, it follows that $\sum_{e\in\mathcal{E}} \int_0^{\ell_e} \frac{\mathrm{d}u_e}{\mathrm{d}x}(x) \frac{\mathrm{d}g_e}{\mathrm{d}x}(x)\mathrm{d}x = \int_\Gamma \frac{\mathrm{d}u}{\mathrm{d}x}\frac{\mathrm{d}g}{\mathrm{d}x}\, \mathrm{d}x$. For the second part, we have

$$\sum_{e\in\mathcal{E}} \left[\frac{\mathrm{d}u_e}{\mathrm{d}x}(x) g_e(x)\right]_0^{\ell_e} = \sum_{e\in\mathcal{E}} \frac{\mathrm{d}u_e}{\mathrm{d}x}(\ell_e) g_e(\ell_e) - \sum_{e\in\mathcal{E}} \frac{\mathrm{d}u_e}{\mathrm{d}x}(0) g_e(0)$$

and further

$$\begin{aligned}
\sum_{e\in\mathcal{E}} \frac{\mathrm{d}u_e}{\mathrm{d}x}(0) g_e(0) &= \sum_{e\in\mathcal{E}} \frac{\mathrm{d}u_e}{\mathrm{d}x}(o(e)) g_e(o(e)) \\
&= \frac{1}{2} \sum_{v\in\mathcal{V}} \sum_{e\in\mathcal{E}_v} \frac{\mathrm{d}u_e}{\mathrm{d}x}(v) g_e(v) = \frac{1}{2} \sum_{v\in\mathcal{V}} g(v) \sum_{e\in\mathcal{E}_v} \frac{\mathrm{d}u_e}{\mathrm{d}x}(v)
\end{aligned}$$

since g is continuous. On the other hand, it holds that

$$\sum_{e\in\mathcal{E}} \frac{\mathrm{d}u_e}{\mathrm{d}x}(\ell_e) g_e(\ell_e) = -\sum_{e\in\mathcal{E}} \frac{\mathrm{d}u_e}{\mathrm{d}x}(t(e)) g_e(t(e)) = -\frac{1}{2} \sum_{v\in\mathcal{V}} \sum_{e\in\mathcal{E}_v} \frac{\mathrm{d}u_e}{\mathrm{d}x}(v) g_e(v).$$

Note that we had to change the sign since the derivative in $\frac{\mathrm{d}u_e}{\mathrm{d}x}(t(e))$ is assumed to be taken away from the vertex $t(e)$, whereas the derivative $\frac{\mathrm{d}u_e}{\mathrm{d}x}(\ell_e)$ is taken in the direction of the coordinate ℓ_e. Together, we conclude that

$$\sum_{e\in\mathcal{E}} \left[\frac{\mathrm{d}u_e}{\mathrm{d}x}(x) g_e(x)\right]_0^{\ell_e} = -\sum_{v\in\mathcal{V}} g(v) \sum_{e\in\mathcal{E}_v} \frac{\mathrm{d}u_e}{\mathrm{d}x}(v)$$

which completes the proof. □

---

[3] Ainsworth, M., Weller, A.: A spectral Galerkin method for reaction-diffusion equations on metric graphs [unpublished manuscript], 2021.

Under Neumann-Kirchhoff conditions, the following lemma is a direct consequence.

Lemma 2.19 *For Neumann-Kirchhoff conditions, the integration by parts formula*

$$(\mathcal{H}u, g)_\Gamma = \mathfrak{h}(u, g) \tag{2.11}$$

holds and the quadratic form of $\mathcal{H}$ *is given by*

$$(\mathcal{H}u, u)_\Gamma = \mathfrak{h}(u, u) = \sum_{e\in\mathcal{E}} \int_e \left(\frac{\mathrm{d}u_e}{\mathrm{d}x}(x)\right)^2 \mathrm{d}x. \tag{2.12}$$

Proof Follows directly from Theorem 2.18 since $\sum_{e\in\mathcal{E}_v} \frac{\mathrm{d}u_e}{\mathrm{d}x}(v) = 0$ for all $v \in \mathcal{V}$. □

It is now straightforward to pose partial differential equations in weak formulation on Γ. Let us start with the introductory elliptic model equations.

Problem 2.20 (Weak Elliptic Differential Equations) Let $f \in L^2(\Gamma)$ and $\rho \in \mathbb{R}^+$. The weak formulation of the elliptic equation $\mathcal{H}u = f$ on Γ is given by:

$$\text{Find } u \in H^1(\Gamma) : \mathfrak{h}(u, g) = (f, g)_\Gamma \quad \text{for all } g \in H^1(\Gamma). \tag{2.13}$$

Equivalently, the weak formulation of $\mathcal{H}u + \rho u = f$ on Γ is given by:

$$\text{Find } u \in H^1(\Gamma) : \mathfrak{h}_\rho(u, g) = (f, g)_\Gamma \quad \text{for all } g \in H^1(\Gamma) \tag{2.14}$$

where $\mathfrak{h}_\rho(u, g) := \mathfrak{h}(u, g) + \rho(u, g)_\Gamma$.

For $\rho > 0$, (2.14) admits a unique weak solution [2]. We will later need problem (2.14) with $\rho = 1$ in which case we use the notation $\mathfrak{h}_1(u, g) := \mathfrak{h}(u, g) + (u, g)_\Gamma$.

We will now turn to weak formulations of the initial boundary value problems (Problems 2.16 and 2.17). Following the convention in [22], we associate with u a mapping from $[0, T] \to L^2(\Gamma)$ or $H^1(\Gamma)$ defined by $[\boldsymbol{u}(t)](x) := u(x, t)$. Moreover, we adopt the notation $\boldsymbol{u}' = \frac{\mathrm{d}}{\mathrm{d}t}\boldsymbol{u}$. We then obtain the following weak formulation.

Problem 2.21 (Weak IBVP for Generalized Heat Equation) Let $\boldsymbol{f} \in L^2([0, T]; L^2(\Gamma))$. The weak formulation of Problem 2.16 consists of finding $\boldsymbol{u} \in L^2([0, T]; H^1(\Gamma))$ with $\boldsymbol{u}' \in L^2([0, T]; L^2(\Gamma))$ such that

$$\left(\boldsymbol{u}', g\right)_\Gamma + \mathfrak{h}(\boldsymbol{u}, g) = (\boldsymbol{f}, g)_\Gamma \tag{2.15}$$

for all $g \in H^1(\Gamma)$ and a.e. time $0 \leq t \leq T$ with $\boldsymbol{u}(0) = u^0$ for $u^0 \in \mathrm{dom}_{\mathcal{H},\mathrm{NK}}$.

Analogous to [22], Section 7.1.2, one can shown that Problem 2.21 has a unique weak solution under the assumption that the bilinear form $\mathfrak{h}(u, g)$ fulfills the following energy estimates for $u, g \in H^1(\Gamma)$:

$$|\mathfrak{h}(u, g)| \leq \gamma_1 \|u\|_{H^1(\Gamma)} \|g\|_{H^1(\Gamma)} \qquad \text{for } \gamma_1 > 0 \tag{2.16}$$

$$\gamma_2 \|u\|^2_{H^1(\Gamma)} \leq \mathfrak{h}(u, u) + \gamma_3 \|u\|^2_\Gamma \qquad \text{for } \gamma_2 > 0, \gamma_3 \geq 0. \tag{2.17}$$

The first condition is the continuity condition familiar from the Lax-Milgram theorem and the second condition is also known as *Gårding's inequality*. It is easy to see that our bilinear form

$$\mathfrak{h}(u, g) = \int_\Gamma \frac{\mathrm{d}u}{\mathrm{d}x} \frac{\mathrm{d}g}{\mathrm{d}x} \mathrm{d}x$$

meets the requirements since

$$\text{(2.16)} \quad |\mathfrak{h}(u, g)| = \left|\int_\Gamma \tfrac{\mathrm{d}u}{\mathrm{d}x} \tfrac{\mathrm{d}g}{\mathrm{d}x} \mathrm{d}x\right| \leq \left\|\tfrac{\mathrm{d}u}{\mathrm{d}x}\right\|_\Gamma \left\|\tfrac{\mathrm{d}g}{\mathrm{d}x}\right\|_\Gamma \leq \|u\|_{H^1(\Gamma)} \|g\|_{H^1(\Gamma)}$$

$$\text{(2.17)} \quad \mathfrak{h}(u, u) = \int_\Gamma \tfrac{\mathrm{d}u}{\mathrm{d}x} \tfrac{\mathrm{d}u}{\mathrm{d}x} \mathrm{d}x = \left\|\tfrac{\mathrm{d}u}{\mathrm{d}x}\right\|^2_\Gamma = \|u\|^2_{H^1(\Gamma)} - \|u\|^2_\Gamma.$$

In particular, (2.17) implies that

$$|||u||| := (\mathfrak{h}(u, u))^{\frac{1}{2}} = \left\|\frac{\mathrm{d}u}{\mathrm{d}x}\right\|_\Gamma = |u|_{H^1(\Gamma)} \tag{2.18}$$

is a semi-norm. Note that the requirement on the initial condition in (2.15) makes sense since one can show that $\boldsymbol{u} \in C([0, T]; L^2(\Gamma))$ (see e.g. , [22], Th. 3, Sec. 5.9.2).

Let us now turn to semilinear parabolic equations.

Problem 2.22 (Weak IBVP for Reaction-Diffusion Equation) Under the assumptions of Problem 2.17, the weak formulation of Problem 2.17 consists of finding
$\boldsymbol{u} \in L^2([0, T]; H^1(\Gamma))$ with $\boldsymbol{u}' \in L^2([0, T]; L^2(\Gamma))$ such that

$$(\boldsymbol{u}', g)_\Gamma + \mathfrak{h}(\boldsymbol{u}, g) = (\mathcal{R}(\boldsymbol{u}), g)_\Gamma \tag{2.19}$$

for all $g \in H^1(\Gamma)$ and a.e. time $0 \leq t \leq T$ with $\boldsymbol{u}(0) = u^0$ for $u^0 \in \mathrm{dom}_{\mathcal{H},\mathrm{NK}}$.

The well-posedness of Problem 2.22 can be shown by a fixed point method, see for example [22], Theorem 2 in Section 9.2.

2.4 Examples and Test Problems

Throughout this work, we will frequently encounter several type of example graphs and test problems, which I will briefly introduce in the present subsection.

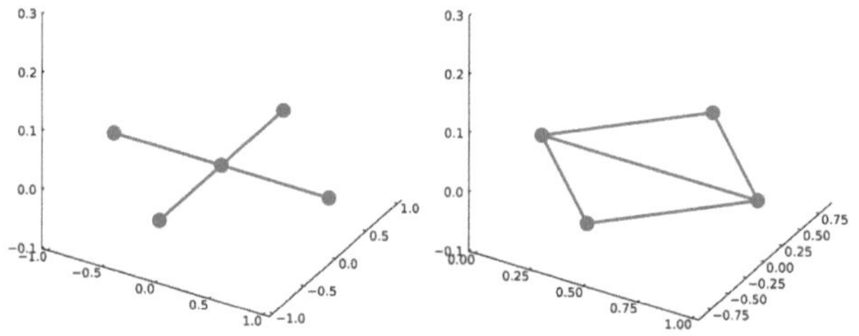

(a) (b)

Fig. 2.3 Illustration of the graphs from Example 2.23. (**a**) Star graph. (**b**) Diamond graph

Example 2.23

(a) A *star graph* Γ_{star} with n vertices has one central vertex that is connected to all other vertices of the graph. The remaining vertices are only connected to the central vertex, thus, a star graph always has $m = n - 1$ edges. If not stated otherwise, we always consider a star graph with $n = 5$ vertices and $m = 4$ edges of equilateral length $\ell = 1$, compare Fig. 2.3a.
(b) A *diamond graph* Γ_{dia} has $n = 4$ vertices and $m = 5$ edges. It arises from a cycle graph by adding one edge in the middle. If not stated otherwise, we will always assume the diamond graph to be equilateral with $\ell = 1$ as illustrated in Fig. 2.3b.

For these graphs, we construct several test problems to test the approximation quality of numerical solutions. The first one is an elliptic boundary value problem on Γ_{star}.[4]

Test Problem 2.20 *Let Γ_{star} be a star graph with $n = 5$ vertices and $m = 4$ edges of equilateral length $\ell = \pi + \frac{\pi}{2}$. Then, the elliptic boundary value problem of finding $u \in \text{dom}_{\mathcal{H},\text{NK}}$ such that*

$$\mathcal{H}u + u = f \text{ on } \Gamma$$

with right-hand side f defined by

$$f_{e_1}(x) = -6\sin(x), \quad f_{e_2} = 2\sin(x), \quad f_{e_3} = 2\sin(x), \quad f_{e_4} = 2\sin(x)$$

for $x \in [0, \pi + \frac{\pi}{2}]$ has the exact solution $u^*(x) = \frac{1}{2}f(x)$.

Similarly, we specify an elliptic test problem for Γ_{dia}.

[4] The formulation of this test problem was the result of a discussion with Dietrich Braess, Angela Kunoth and Max Brockmann in January 2023.

Test Problem 2.21 *Let Γ_{dia} be a simple graph with $n = 4$ vertices and $m = 5$ edges, each of length $\ell = 2\pi$. Then, the elliptic boundary value problem of finding $u \in \mathrm{dom}_{\mathcal{H},\mathrm{NK}}$ such that*

$$\mathcal{H}u + u = f \text{ on } \Gamma$$

with right-hand side f defined by

$$f_{e_1}(x) = 2\sin(x), \quad f_{e_2} = -4\sin(x), \quad f_{e_3} = 2\sin(x),$$
$$f_{e_4} = 2\sin(x), \quad f_{e_5} = -2\sin(x)$$

for $x \in [0, 2\pi]$ has the exact solution $u^*(x) = \frac{1}{2}f(x)$.

The solution of both test problems on the underlying metric graph is illustrated in Fig. 2.4. Clearly, we constructed the test problems in such a way that the solution fulfills the Neumann-Kirchhoff conditions.

In order to formulate test problems for linear parabolic equations, we will choose a solution of the eigenvalue equation (2.5) on the metric graph as initial condition. As in the one dimensional setting, the solution of the PDE can then be obtained explicitly with a separation of variables approach.

Test Problem 2.22 *Let Γ_{star} be a star graph with $n = 5$ vertices and $m = 4$ edges, but here each of length $\ell = 1$, and let $T \in \mathbb{R}^+$. The presented test problem consists of the IBVP*

$$\frac{\partial u}{\partial t} + \mathcal{H}u = 0 \quad \text{on } \Gamma \times [0, T]$$
$$u(0) = u^0 \quad \text{for } x \in \Gamma$$

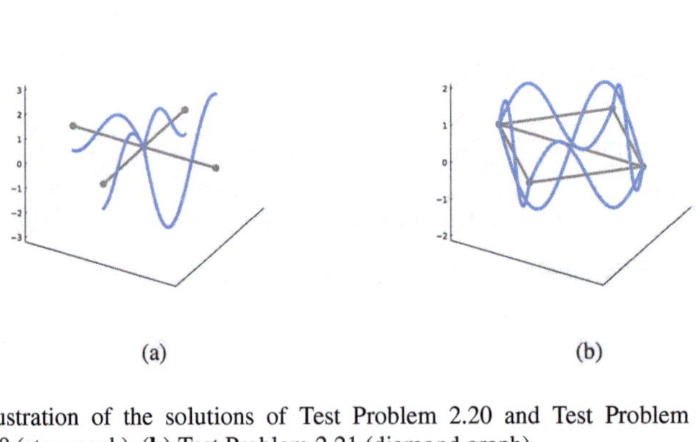

(a) (b)

Fig. 2.4 Illustration of the solutions of Test Problem 2.20 and Test Problem 2.21. (**a**) Test Problem 2.20 (star graph). (**b**) Test Problem 2.21 (diamond graph)

Table 2.1 Coefficients for the initial condition on Γ_{star} (Test Problem 2.22)

	e_1	e_2	e_3	e_4
A_e	0.707107	0.707107	0.707107	0.707107
B_e	0	0	0	0

Table 2.2 Coefficients for the initial condition on Γ_{dia} (Test Problem 2.21)

	e_1	e_2	e_3	e_4	e_5
A_e	0.57735	0.57735	0.57735	0	−0.57735
B_e	0.204124	−0.408248	0.204124	−0.612372	−0.204124

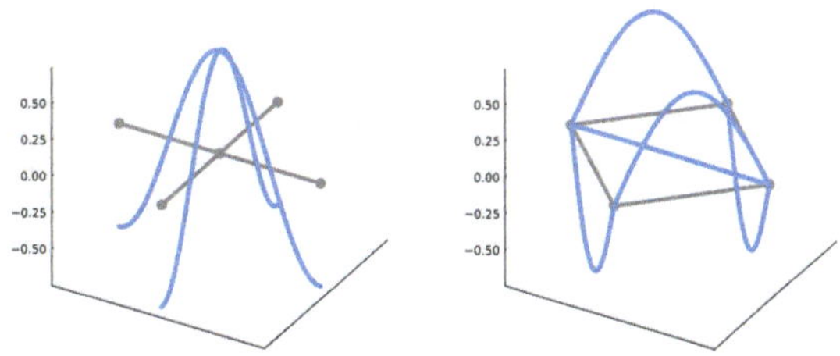

Fig. 2.5 Initial conditions for Test Problem 2.22 and Test Problem 2.21

under Neumann-Kirchhoff conditions and with an initial condition of the form

$$u_e^0(x) = A_e \cos\left(\sqrt{\lambda}x\right) + B_e \sin\left(\sqrt{\lambda}x\right) \tag{2.20}$$

where $\lambda = \pi^2$ *and the constants* A_e, B_e *are given in Table 2.1. This initial condition is an eigenfunction of* Γ *and thereby the exact solution is given by* $u^*(x,t) = \exp(-t\lambda)\, u^0(x)$.

Test Problem 2.21 *Consider the IBVP from Test Problem 2.22 posed on the diamond graph* Γ_{dia} *with* $n = 4$ *vertices and* $m = 5$ *edges, each of length* $\ell = 1$. *The initial condition is of the form* (2.20) *with* $\lambda = 3.65\text{e}{+}00$ *and coefficients given in Table 2.2.*

A visualization of the two initial condition can be found in Fig. 2.5.

We will moreover consider large scale random graphs constructed by the Barabási-Albert [4] or Erdős-Rényi [21] model.

Example 2.24

(a) A Barabási-Albert graph Γ_{BA} represents a scale-free network constructed following a preferential attachment model. The model starts with a small number of vertices and subsequently adds new vertices that are connected to the existing vertices with d edges. In doing so, the probability that an edge will be added between the new vertex and the existing vertices depends on the degree of the existing vertices (the new vertices are preferentially connected to existing high-degree vertices). The resulting graph with n vertices is a scale-free network, i.e., the degree distribution follows a power law. We will typically consider sparse graphs with few edges and thus choose $d = 2$ or $d = 3$ in our experiments.

(b) An Erdős-Rényi graph Γ_{ER} is a graph with n vertices where each edge between two vertices is added with a probability p.

The last two examples originate from [2]. These are examples of non-trivial graphs which however still can be plotted in a plane, thereby allowing it to nicely visualize functions on them.

Example 2.25

(a) We define the so-called *graphene graph* Γ_{graphene} by connecting two 6-cycles with one edge. The resulting graph has $n = 12$ vertices and $m = 13$ edges and is illustrated in Fig. 2.6 for equilateral edge length $\ell = 1$.

(b) Γ_{tree} is a *tree graph* with $n = 16$ vertices and $m = 15$ edges of length $\ell = 1$ as given in Fig. 2.6. The first edge of the tree is also referred to as the *stem*.

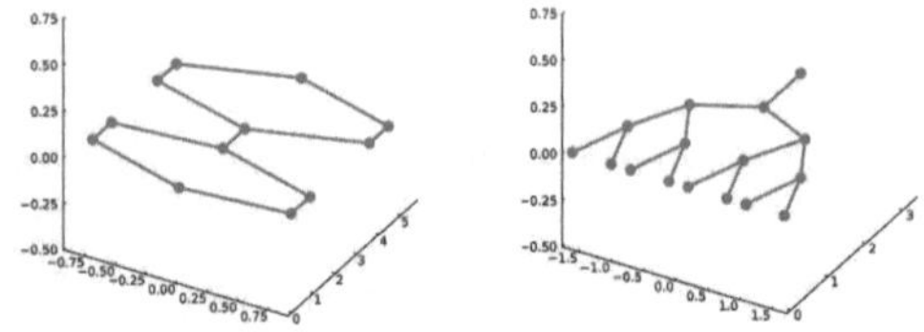

Fig. 2.6 Graphene and tree graph with equilateral edge length $\ell = 1$

Chapter 3
Finite Element Method

Abstract This chapter derives the finite element approach proposed by Arioli and Benzi (IMA J Numer Anal 38(3):1119–1163, 2018) and its application to the semidiscretization of parabolic equations. The discretization of metric graphs leading to the concept of *extended graphs* is reviewed to allow a structured representation of the finite element semidiscretization. The main new aspect elaborated for this book is the solution of the semidiscretized systems with implicit-explicit time stepping methods combined with a multigrid ansatz for the solution of the arising linear systems of equations. Moreover, an $L^2(\Gamma)$-error estimate will be derived which allows to use standard theory of Galerkin finite element methods for parabolic problems (see Thomée, Galerkin finite element methods for parabolic problems. Springer, 1997) to deduce estimates for the error between the solutions of the semidiscrete and the continuous problem.

3.1 Discretization and Extended Graphs

The finite element approach requires a preliminary *discretization* of the domain, i.e., the metric graph in consideration. In this context, [2] introduce the concept of *extended graphs* where the discretization of Γ is itself understood as a metric graph $\tilde{\Gamma}$ by considering the discretization points as additional vertices. This extended graph will turn out to be extremely useful since its graph matrices allow a structured representation of the finite element semidiscretization.

Before we turn to the finite element approximation, we will therefore give a detailed discussion of $\tilde{\Gamma}$, introduce a weighted form of its graph matrices and review their efficient computation. Clearly, the motivation and ideas of this exposition and the techniques used arose from [2] as mentioned above, yet I tried to present them in a generalized form detached from the finite element setting.

A. Weller, *Numerical Methods for Metric Graphs*, Lecture Notes in Mathematics 2382, https://doi.org/10.1007/978-3-032-05011-3_3

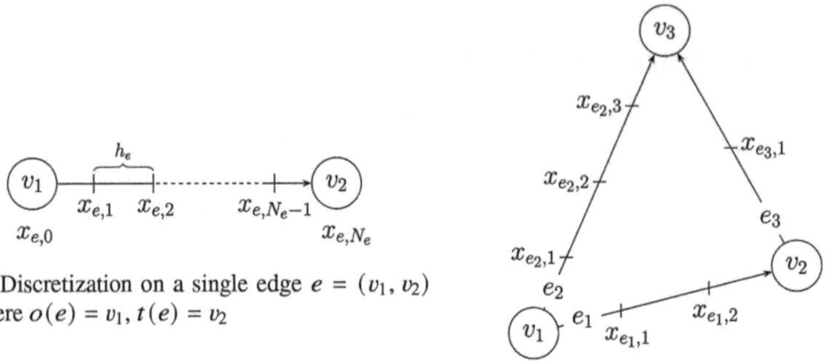

(a) Discretization on a single edge $e = (v_1, v_2)$ where $o(e) = v_1, t(e) = v_2$

(b) Discretization of a simple graph with different step sizes per edge

Fig. 3.1 Metric graph discretization for a single edge and a simple example graph. (**a**) Discretization on a single edge $e = (v_1, v_2)$ where $o(e) = v_1, t(e) = v_2$. (**b**) Discretization of a simple graph with different step sizes per edge

3.1.1 Discretization

Let Γ be a metric graph with an arbitrary but fixed orientation on the edges. For each edge $e \in \mathcal{E}$ with length ℓ_e, we discretize the domain $[0, \ell_e]$ by

$$x_{e,k} = k \times h_e, \quad k = 0, \ldots, N_e$$

where k increases in the direction of e, h_e is the step size chosen on edge e, and $N_e + 1$ is the number of discretization points this choice delivers. Note that the first discretization point $x_{e,0} = 0$ is placed at the start vertex $o(e)$ and the last discretization point $x_{e,N_e} = \ell_e$ at the end vertex $t(e)$. We will often distinguish between *vertex grid points* $x_{e,0}$, x_{e,N_e} and *inner grid points* $x_{e,k}$ for $k = 1, \ldots, N_e - 1$. The discretization is illustrated in Fig. 3.1.

In particular, it is important to observe the different lengths ℓ_e of the edges. In general, it is therefore not possible to choose the same step size h_e on each edge. In the special case of equilateral graphs, we choose the simplified notation h and N since we will then apply a discretization with a uniform step size h on each edge.

3.1.2 Extended Graphs

The inner grid points $x_{e,k}$ on the edges of Γ can be understood as additional vertices. The resulting *extended graph* was introduced by [2] and will be referred to as $\tilde{\Gamma}$. In this context, we will denote the inner grid points, understood as vertices, by $v_{e,k}$ for $e \in \mathcal{E}$ and $k = 1, \ldots, N_e - 1$.

Definition 3.1 For a metric graph Γ with vertex set $\mathcal{V}$, edge set $\mathcal{E}$ and edge lengths $\boldsymbol{\ell}$, we define the *extended graph* $\tilde{\Gamma}$ with vertex set $\tilde{\mathcal{V}}$, edge set $\tilde{\mathcal{E}}$ and step sizes $\boldsymbol{h}$ as follows.

(a) The vertex set of $\tilde{\Gamma}$ is given by

$$\tilde{\mathcal{V}} := \mathcal{V} \cup \bigcup_{e\in\mathcal{E}} \{v_{e,k}, k = 1, \dots, N_e - 1\}$$

and contains $|\tilde{\mathcal{V}}| = n + \sum_{e\in\mathcal{E}}(N_e - 1) =: \tilde{n}$ elements.

(b) For each original edge e, the extended graph has N_e partitioned edges

$$\tilde{\mathcal{E}}_e := \big(o(e), v_{e,1}\big) \cup \bigcup_{k=1}^{N_e-2} (v_{e,k}, v_{e,k+1}) \cup (v_{e,N_e-1}, t(e)).$$

Together, we define $\tilde{\mathcal{E}} := \bigcup_{e\in\mathcal{E}} \tilde{\mathcal{E}}_e$. In total, $\tilde{\Gamma}$ has $|\tilde{\mathcal{E}}| = \sum_{e\in\mathcal{E}} N_e =: \tilde{m}$ edges.

(c) Induced by the step size, each edge $\tilde{e} \in \tilde{\mathcal{E}}_e$ has length h_e. All the step sizes chosen on the different edges are collected in $\boldsymbol{h} := (h_{e_1}, \dots, h_{e_m})^T \in \mathbb{R}^m$.

Note that we still identify the vertices with the corresponding grid points and only adjust the notation to emphasize that the grid can be interpreted as a new graph.

Remark 3.2 If we want to emphasize from which step sizes $\tilde{\Gamma}$ arises, we will work with a subscript, for instance $\tilde{\Gamma}_{\boldsymbol{h}}, \tilde{\mathcal{V}}_{\boldsymbol{h}}, \dots$. However, if $\boldsymbol{h}$ is fixed or clear from the context, we will prefer the shorthand notation without subscript to improve readability.

To facilitate the exposition, we will deviate from the usual numbering of the edges (from smaller to larger vertex index) and instead number and orient the extended edges $\tilde{e}$ sequentially according to their associated original edge, as illustrated in Fig. 3.2.

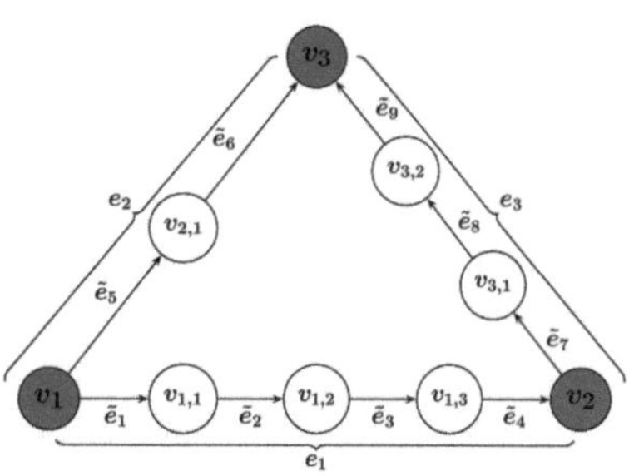

Fig. 3.2 Edge numeration and orientation in the extended graph

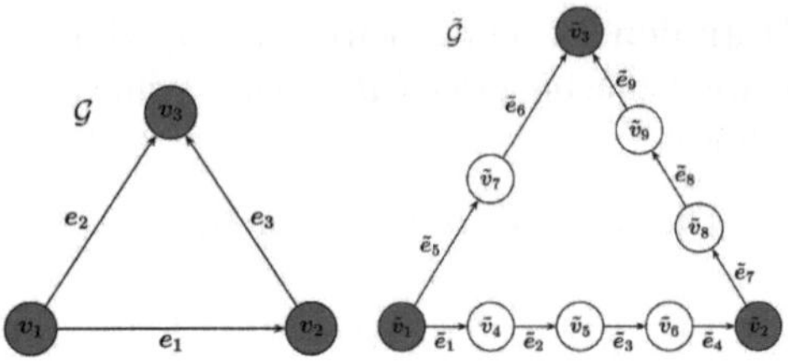

Fig. 3.3 Vertex numeration in the extended combinatorial graph

Definition 3.3 We define the *underlying combinatorial extended graph* $\tilde{\mathcal{G}}$ by $\tilde{\mathcal{G}} = (\tilde{\mathcal{V}}, \tilde{\mathcal{E}})$ with *extended graph Laplacian matrix*

$$\tilde{\mathbf{L}} := \tilde{\mathbf{D}} - \tilde{\mathbf{A}} = \tilde{\mathbf{N}}\tilde{\mathbf{N}}^T$$

where $\tilde{\mathbf{D}}$, $\tilde{\mathbf{A}}$ and $\tilde{\mathbf{N}}$ are the degree, adjacency and incidence matrix of $\tilde{\mathcal{G}}$.

In this context, we assume the following ordering of the vertices in $\tilde{\mathcal{G}}$: We start with the original vertices $v_1, \ldots, v_n \in \mathcal{V}$ according to their numbering in the original graph $\mathcal{G}$. Next, we enumerate the inner vertices on the first edge e_1 by $n+1, \ldots, n+N_{e_1}-1$ followed by the inner vertices on the second edge e_2 and so on, compare Fig. 3.3.

Respecting the introduced ordering, the graph Laplacian matrix of $\tilde{\mathcal{G}}$ has the block structure

$$\tilde{\mathbf{L}} = \begin{bmatrix} \tilde{\mathbf{L}}_{\mathcal{V}\mathcal{V}} & \tilde{\mathbf{L}}_{\mathcal{V}\mathcal{E}} \\ \tilde{\mathbf{L}}_{\mathcal{E}\mathcal{V}} & \tilde{\mathbf{L}}_{\mathcal{E}\mathcal{E}} \end{bmatrix}.$$

The subscripts emphasize that the blocks reflect

1. the adjacency between and the degree of the original vertices ($\tilde{\mathbf{L}}_{\mathcal{V}\mathcal{V}}$),
2. the adjacency between and the degree of the inner vertices ($\tilde{\mathbf{L}}_{\mathcal{E}\mathcal{E}}$) and
3. the adjacency between the original vertices and inner vertices ($\tilde{\mathbf{L}}_{\mathcal{V}\mathcal{E}}$)

in the extended graph. In particular, $\tilde{\mathbf{L}}$ is symmetric, i.e., $\tilde{\mathbf{L}}_{\mathcal{V}\mathcal{E}}^T = \tilde{\mathbf{L}}_{\mathcal{E}\mathcal{V}}$. If we assume that there is at least one grid point per edge, we obtain

$$\tilde{\mathbf{L}}_{\mathcal{V}\mathcal{V}} = \mathbf{D} \in \mathbb{R}^{n \times n}$$

since then $(v_i, v_j) \notin \tilde{\mathcal{E}}$ for all original vertices $v_i, v_j \in \mathcal{V}$. Moreover, the block $\tilde{\mathbf{L}}_{\mathcal{E}\mathcal{E}}$ itself is of the block structure

$$\tilde{\mathbf{L}}_{\mathcal{E}\mathcal{E}} = \text{blkDiag}(\tilde{\mathbf{L}}_{e_1}, \ldots, \tilde{\mathbf{L}}_{e_m}) := \begin{bmatrix} \tilde{\mathbf{L}}_{e_1} & & \\ & \ddots & \\ & & \tilde{\mathbf{L}}_{e_m} \end{bmatrix}.$$

Each block $\tilde{\mathbf{L}}_e$ corresponds to the inner vertices on edge $e \in \mathcal{E}$ and is of the form

$$\tilde{\mathbf{L}}_e = \begin{pmatrix} 2 & -1 & & & \\ -1 & 2 & -1 & & \\ & \ddots & \ddots & \ddots & \\ & & -1 & 2 & -1 \\ & & & -1 & 2 \end{pmatrix} \in \mathbb{R}^{(N_e-1)\times(N_e-1)}.$$

Remark 3.4 We point out to the reader familiar with [2] that the ordering of the blocks in $\tilde{\mathbf{L}}$ depends on the ordering of the vertices in the extended graph and therefore deviates from [2] where the internal discretization points are enumerated first.

Of special interest for the representation of the finite element discretization matrices will be a weighted version of the extended graph Laplacian matrix that accounts for possible different step sizes on the edges.

Definition 3.5 We equip the edges of the extended graph with a weight according to the reciprocal of their length, i.e., $w_{\tilde{e}} = 1/h_e$ for all $\tilde{e} \in \tilde{\mathcal{E}}_e$.

(a) The *weighted extended graph Laplacian matrix* is defined by $\hat{\mathbf{L}} := \hat{\mathbf{D}} - \hat{\mathbf{A}}$ where the weighted version of the adjacency matrix is defined by

$$(\hat{\mathbf{A}})_{i,j} := \begin{cases} w_{(\tilde{v}_i,\tilde{v}_j)}, & \text{if } (\tilde{v}_i,\tilde{v}_j) \in \tilde{\mathcal{E}} \\ 0 & \text{otherwise} \end{cases}$$

and the diagonal weighted degree matrix has entries $(\hat{\mathbf{D}})_{i,i} := \sum_{\tilde{e}\in\tilde{\mathcal{E}}_{\tilde{v}_i}} w_{\tilde{e}}$.

(b) If we define the *edge-weight matrix* $\tilde{\mathbf{W}} \in \mathbb{R}^{\tilde{m}\times\tilde{m}}$ by $\tilde{\mathbf{W}} := \operatorname{diag}((w_{\tilde{e}})_{\tilde{e}\in\tilde{\mathcal{E}}})$, the weighted extended graph Laplacian matrix can be expressed in terms of the incidence matrix of the extended graph $\tilde{\mathbf{N}} \in \mathbb{R}^{\tilde{n}\times\tilde{m}}$ as

$$\hat{\mathbf{L}} := \tilde{\mathbf{N}}\tilde{\mathbf{W}}\tilde{\mathbf{N}}^T.$$

In particular, following the chosen edge numbering, the weight matrix is given by

$$\tilde{\mathbf{W}} = \operatorname{blkDiag}\left(\left\{\frac{1}{h_e}\mathbf{I}_{N_e}\right\}_{e\in\mathcal{E}}\right)$$

where $\mathbf{I}_{N_e}$ denotes the identity matrix of size N_e.

Clearly, the observations on the block structure also apply to the weighted version

$$\hat{\mathbf{L}} = \begin{pmatrix} \hat{\mathbf{L}}_{\mathcal{V}\mathcal{V}} & \hat{\mathbf{L}}_{\mathcal{V}\mathcal{E}} \\ \hat{\mathbf{L}}_{\mathcal{E}\mathcal{V}} & \hat{\mathbf{L}}_{\mathcal{E}\mathcal{E}} \end{pmatrix}.$$

Now we have

$$\hat{\mathbf{L}}_{\mathcal{V}\mathcal{V}} = \operatorname{diag}\left(\left\{\sum_{e\in\mathcal{E}_v}\frac{1}{h_e}\right\}_{v\in\mathcal{V}}\right)$$

and $\hat{\mathbf{L}}_{\mathcal{E}\mathcal{E}} = \operatorname{blkDiag}(\hat{\mathbf{L}}_{e_1}, \ldots, \hat{\mathbf{L}}_{e_m})$ with

$$\hat{\mathbf{L}}_e = \frac{1}{h_e}\begin{pmatrix} 2 & -1 & & & \\ -1 & 2 & -1 & & \\ & \ddots & \ddots & \ddots & \\ & & -1 & 2 & -1 \\ & & & -1 & 2 \end{pmatrix}.$$

3.1.3 Extended Graph Construction

A straightforward approach to directly compute the extended graph is the explicit expansion of the vertex and edge set by inserting inner grid points and edges. Yet, this process is computationally expensive, especially for small step sizes. Moreover, for later theoretical investigations, it will turn out to be useful to construct the extended graph matrices by some manipulations of the incidence matrix $\mathbf{N}$ of the original graph as developed in [2]. For the remainder of this subsection, we will therefore briefly demonstrate this derivation of $\tilde{\mathbf{N}}$ as seen in [2].

We first separate the extended incidence matrix $\tilde{\mathbf{N}}$ in two parts:

$$\tilde{\mathbf{N}} = \begin{pmatrix} \tilde{\mathbf{N}}_{\mathcal{V}} \\ \tilde{\mathbf{N}}_{\mathcal{E}} \end{pmatrix} = \begin{array}{c} \\ \tilde{v}_1 \\ \vdots \\ \tilde{v}_n \\ \tilde{v}_{n+1} \\ \vdots \\ \tilde{v}_{\tilde{n}} \end{array} \begin{array}{c} \tilde{e}_1\ \tilde{e}_2\ \ldots\ \tilde{e}_{\tilde{m}} \\ \boxed{\tilde{\mathbf{N}}_{\mathcal{V}}} \\ \boxed{\tilde{\mathbf{N}}_{\mathcal{E}}} \end{array}.$$

Here, $\tilde{\mathbf{N}}_{\mathcal{V}} \in \mathbb{R}^{n\times\tilde{m}}$ represents the edges incident to the original vertices and $\tilde{\mathbf{N}}_{\mathcal{E}} \in \mathbb{R}^{\tilde{n}-n} \times \tilde{m}$ the edges incident to the inner vertices. The appearance of the latter can be easily specified since the inner vertices represent a path graph on each edge. Thus, $\tilde{\mathbf{N}}_{\mathcal{E}}$ is a block diagonal matrix

$$\tilde{\mathbf{N}}_{\mathcal{E}} = \begin{bmatrix} \tilde{\mathbf{N}}_{e_1} & & \\ & \ddots & \\ & & \tilde{\mathbf{N}}_{e_m} \end{bmatrix}$$

with each block $\tilde{\mathbf{N}}_e \in \mathbb{R}^{(N_e-1)\times N_e}$ given by

$$\tilde{\mathbf{N}}_e = \begin{pmatrix} 1 & -1 & & & \\ & 1 & -1 & & \\ & & \ddots & \ddots & \\ & & & 1 & -1 \end{pmatrix}.$$

The incidence between the original vertices and the extended graph edges can be represented by horizontally concatenated matrices as

$$\tilde{\mathbf{N}}_{\mathcal{V}} = \left[\tilde{\mathbf{N}}_{\mathcal{V}_{e_1}}, \ldots, \tilde{\mathbf{N}}_{\mathcal{V}_{e_m}}\right].$$

Again, each block $\tilde{\mathbf{N}}_{\mathcal{V}_e} \in \mathbb{R}^{n\times N_e}$ corresponds to an edge of the original graph. In other words, the block $\tilde{\mathbf{N}}_{\mathcal{V}_e}$ contains the information about the incidence of the original vertices to the subdivisions $\tilde{e} \in \tilde{\mathcal{E}}_e$. Clearly, only the first or the last edge partition can be attached to an original vertex. The blocks can therefore be specified by expanding the original incidence matrix $\mathbf{N}$ as illustrated in Fig. 3.4.

To do so, we must first differentiate between ingoing and outgoing edges in $\mathbf{N}$. We therefore define

$$\mathbf{N}^{\text{in}} := \frac{1}{2}(\mathbf{N} + |\mathbf{N}|) \quad \text{and} \quad \mathbf{N}^{\text{out}} := \frac{1}{2}(\mathbf{N} - |\mathbf{N}|)$$

where $|\mathbf{N}|$ denotes the entry-wise absolute values of $\mathbf{N}$. The matrices $\tilde{\mathbf{N}}_{\mathcal{V}_e}$ can be computed as

$$\tilde{\mathbf{N}}_{\mathcal{V}_e} = \mathbf{N}_e^{\text{out}} \otimes (\mathbf{e}_1^{N_e})^T + \mathbf{N}_e^{\text{in}} \otimes (\mathbf{e}_{N_e}^{N_e})^T$$

where $\mathbf{e}_1^{N_e}$ and $\mathbf{e}_{N_e}^{N_e}$ are defined as the first respectively last column of the identity matrix of size N_e and $\mathbf{N}_e$ denotes the column of $\mathbf{N}$ corresponding to the edge e. With this considerations in place, we summarize the computation of $\tilde{\mathbf{N}}$ in Algorithm 1.

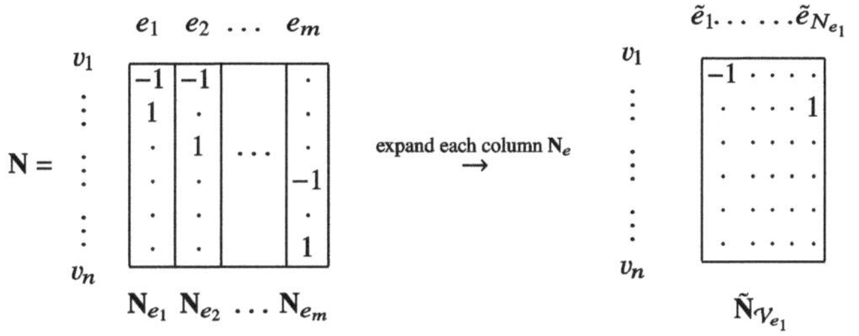

Fig. 3.4 Construction of $\tilde{\mathbf{N}}_{\mathcal{V}} = \left[\tilde{\mathbf{N}}_{\mathcal{V}_{e_1}}, \ldots, \tilde{\mathbf{N}}_{\mathcal{V}_{e_m}}\right]$ from $\mathbf{N}$

Algorithm 1 Extended graph incidence matrix

Input: n, m : number of nodes and edges
$\mathbf{N}$: incidence matrix of original graph
$(N_{e_1}, \ldots, N_{e_m})^T$: vector with inner grid points per edge
Output: extended graph incidence matrix $\tilde{\mathbf{N}}$

▷ construct $\tilde{\mathbf{N}}_{\mathcal{V}}$

Compute $\mathbf{N}^{\text{in}} = \frac{1}{2}(\mathbf{N} + |\mathbf{N}|)$, $\mathbf{N}^{\text{out}} = \frac{1}{2}(\mathbf{N} - |\mathbf{N}|)$
Decompose $\mathbf{N}^{\text{in}} = [\mathbf{N}^{\text{in}}_{e_1}, \ldots, \mathbf{N}^{\text{in}}_{e_m}]$, $\mathbf{N}^{\text{out}} = [\mathbf{N}^{\text{out}}_{e_1}, \ldots, \mathbf{N}^{\text{out}}_{e_m}]$
for $e = e_1, \ldots, e_m$ **do**
  $\mathbf{e}_1 = (1, 0 \ldots, 0)^T \in \mathbb{R}^{N_e}$, $\mathbf{e}_{N_e} = (0, \ldots, 0, 1)^T \in \mathbb{R}^{N_e}$
  $\tilde{\mathbf{N}}_{\mathcal{V}_e} = \mathbf{N}^{\text{out}}_e \otimes (\mathbf{e}^{N_e}_1)^T + \mathbf{N}^{\text{in}}_e \otimes (\mathbf{e}^{N_e}_{N_e})^T$
end for
Set $\tilde{\mathbf{N}}_{\mathcal{V}} = \left[\tilde{\mathbf{N}}_{\mathcal{V}_{e_1}}, \ldots, \tilde{\mathbf{N}}_{\mathcal{V}_{em}}\right]$
for $e = e_1, \ldots, e_m$ **do** ▷ construct $\tilde{\mathbf{N}}_{\mathcal{E}}$
  Compute $\tilde{\mathbf{N}}_e = \text{Tridiagonal}(\text{values} = (0, 1, -1), \text{size} = ((N_e - 1) \times N_e))$
end for
Set $\tilde{\mathbf{N}}_{\mathcal{E}} = \text{blkDiag}(\tilde{\mathbf{N}}_{e_1}, \ldots, \tilde{\mathbf{N}}_{e_m})$, $\tilde{\mathbf{N}} = \begin{pmatrix} \tilde{\mathbf{N}}_{\mathcal{V}} \\ \tilde{\mathbf{N}}_{\mathcal{E}} \end{pmatrix}$ ▷ assemble $\tilde{\mathbf{N}}$

3.2 Finite Element Approximation

As mentioned in the introduction, a finite element method for the spatial discretization has been extensively discussed in [2] where, in particular, the extended graph was introduced to represent the coefficient matrices of the discretization. Therefore, this section is mainly a summary of the elaborations in [2] embedded in the presented notation and with the objective to derive a finite element semidiscretization of the (semi)linear parabolic equations.

Moreover, the $H^1(\Gamma)$-error estimate derived in [2] for elliptic problems will be enhanced to an $L^2(\Gamma)$-error estimate. We may then apply the results in [50] to estimate the error between the solutions of the semidiscrete and the continuous problem.

3.2.1 *Finite Elements*

As elaborated in Sect. 3.1.1, we discretize each edge of a metric graph Γ with step size h_e resulting in N_e subintervals of the form $[x_{e,k}, x_{e,k+1}]$ for $k = 0, \ldots, N_e - 1$. If $x_{e,k}$ is an inner grid point (i.e., $k = 1, \ldots, N_e - 1$), we define the standard hat basis functions as

$$\psi_{e,k}(x) := \begin{cases} 1 - \frac{x_{e,k} - x}{h_e} & \text{if } x \in [x_{e,k-1}, x_{e,k}], \\ 1 + \frac{x_{e,k} - x}{h_e} & \text{if } x \in [x_{e,k}, x_{e,k+1}], \\ 0 & \text{otherwise,} \end{cases} \tag{3.1}$$

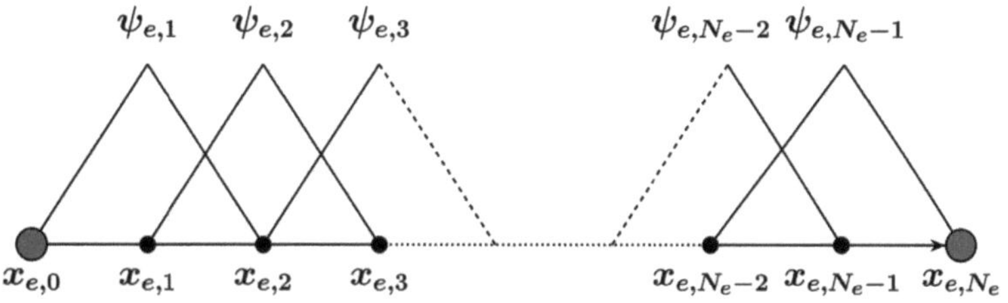

Fig. 3.5 Hat basis functions on an edge e

see Fig. 3.5. Here, $x_{e,0} = o(e)$ and $x_{e,N_e} = t(e)$ as usual.

The hat functions $\psi_{e,k}$ form a basis for

$$V_{h_e} := \{u \in H_0^1(e) : u|_{[x_{e,k},x_{e,k+1}]} \text{ is linear for } k = 0, \ldots, N_e - 1\}. \tag{3.2}$$

To define basis functions on the original vertices, we first introduce the *neighborhood*

$$\mathscr{W}_v := \bigcup_{e \in \mathcal{E}_v^{\text{out}}} \left[o(e), x_{e,1}\right] \cup \bigcup_{e \in \mathcal{E}_v^{\text{in}}} \left[x_{e,N_e-1}, t(e)\right]$$

of a vertex v. Recall that $\mathcal{E}_v^{\text{out}}$ is the set of edges with $v = o(e)$ and $\mathcal{E}_v^{\text{in}}$ with $v = t(v)$, respectively. Thus, $\mathscr{W}_v$ is the metric graph induced by v and its neighboring inner grid points. In particular, the edge parametrizations of $\mathscr{W}_v$ are given by

$$\mathscr{W}_v \cap e = \begin{cases} \left[o(e), x_{e,1}\right] = \left[0, h_e\right] & \text{if } e \in \mathcal{E}_v^{\text{out}}, \\ \left[x_{e,N_e-1}, t(e)\right] = \left[\ell_e - h_e, \ell_e\right] & \text{if } e \in \mathcal{E}_v^{\text{in}}. \end{cases}$$

We may now define basis functions ψ_v in the neighborhood of v by

$$\psi_v(x) = \begin{cases} 1 - \frac{x}{h_e} & \text{if } x \in \mathscr{W}_v \cap e \text{ and } e \in \mathcal{E}_v^{\text{out}}, \\ 1 - \frac{\ell_e - x}{h_e} & \text{if } x \in \mathscr{W}_v \cap e \text{ and } e \in \mathcal{E}_v^{\text{in}}, \\ 0 & \text{otherwise.} \end{cases} \tag{3.3}$$

An example for ψ_v on $\mathscr{W}_v$ with three incident edges is illustrated in Fig. 3.6.

Together with (3.2), we construct the space

$$V_{\boldsymbol{h}}(\Gamma) := \bigoplus_{e \in \mathcal{E}} V_{h_e} \oplus \operatorname{span}\{\psi_v : v \in \mathcal{V}\} \subset H^1(\Gamma). \tag{3.4}$$

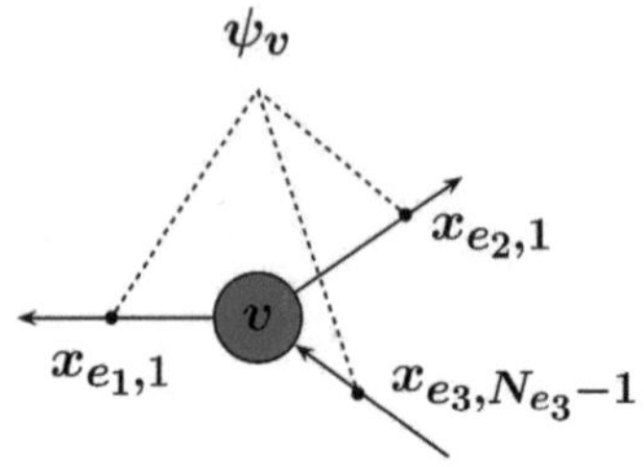

Fig. 3.6 Hat basis functions on $\mathscr{W}_v$

The elements of $V_{\boldsymbol{h}}(\Gamma)$ can be expressed as linear combination of the hat functions, i.e.,

$$u_{\boldsymbol{h}}(x) = \sum_{v \in \mathcal{V}} u_v \, \psi_v(x) + \sum_{e \in \mathcal{E}} \sum_{k=1}^{N_e - 1} u_{e,k} \, \psi_{e,k}(x) \tag{3.5}$$

with coefficients $u_v, u_{e,k} \in \mathbb{R}$.

3.2.2 Semidiscretized System

The weak form of the semilinear parabolic initial boundary value problem (Problem 2.16) is given in Problem 2.21. A Galerkin approximation on $V_{\boldsymbol{h}}$ thus yields the finite element approximation

$$\begin{aligned} \left(\boldsymbol{u}_{\boldsymbol{h}}'(t), g_{\boldsymbol{h}}\right)_\Gamma + \mathfrak{h}(\boldsymbol{u}_{\boldsymbol{h}}(t)', g_{\boldsymbol{h}}) &= (\boldsymbol{f}(t), g_{\boldsymbol{h}})_\Gamma \quad \text{for all} \quad g_{\boldsymbol{h}} \in V_{\boldsymbol{h}}, t > 0 \\ \boldsymbol{u}_{\boldsymbol{h}}(0) &= u_{\boldsymbol{h}}^0 \end{aligned} \tag{3.6}$$

where $u_{\boldsymbol{h}}^0$ is an approximation of the initial condition. In this subsection, we want to derive the stiffness matrix representing the discretization of $\mathfrak{h}(\cdot, \cdot)$ and the mass matrix arising from $(\boldsymbol{u}_{\boldsymbol{h}}, g_{\boldsymbol{h}})_\Gamma$.

Theorem 3.6 *Let Γ be a metric graph and $\tilde{\Gamma}$ the extended graph arising from a discretization with step sizes* $\mathbf{h}$ *(Definition 3.1) and equipped with edge weights $w_{\tilde{e}} = 1/h_e$ for all $\tilde{e} \in \tilde{\mathcal{E}}_e$. Let further*

$$\mathbf{u}(t) := \begin{bmatrix} \mathbf{u}_{\mathcal{V}}(t) \\ \mathbf{u}_{\mathcal{E}}(t) \end{bmatrix}$$

with

$$\mathbf{u}_{\mathcal{V}}(t) := (u_1(t), \ldots, u_n(t))^T,$$

$$\mathbf{u}_{\mathcal{E}}(t) := (u_{e_1,1}(t), \ldots u_{e_1,N_{e_1}-1}(t), \ldots, u_{e_m,1}(t), \ldots, u_{e_m,N_{em}-1}(t))^T$$

be the vector collecting the coefficients of the linear combination in (3.5)*,* $\hat{\mathbf{L}}$ *denote the weighted graph Laplacian of the extended graph and* $\tilde{\mathbf{W}}$ *the edge weight matrix of* $\tilde{\Gamma}$ *as introduced in Definition 3.5. Then, the finite element semidiscretization (3.6) leads to the initial value problem (IVP)*

$$\frac{\mathrm{d}}{\mathrm{d}t}\hat{\mathbf{M}}\mathbf{u}(t) + \hat{\mathbf{L}}\mathbf{u}(t) = \hat{\mathbf{f}}, \quad \mathbf{u}(0) = \mathbf{u}^0 \tag{3.7}$$

where the mass matrix is given by

$$\hat{\mathbf{M}} := \frac{1}{6}\left(|\tilde{\mathbf{N}}\tilde{\mathbf{W}}^{-1}\tilde{\mathbf{N}}^T| + \mathrm{diag}\left(\{(|\tilde{\mathbf{N}}\tilde{\mathbf{W}}^{-1}\tilde{\mathbf{N}}^T|)_{i,i}\}_{i=1}^{\tilde{n}}\right)\right)$$

and

$$\hat{\mathbf{f}}(t) := \begin{bmatrix} \hat{\mathbf{f}}_{\mathcal{V}}(t) \\ \hat{\mathbf{f}}_{\mathcal{E}}(t) \end{bmatrix} \text{ where } \hat{\mathbf{f}}_{\mathcal{E}}(t) = \begin{bmatrix} \hat{\mathbf{f}}_{e_1}(t) \\ \vdots \\ \hat{\mathbf{f}}_{em}(t) \end{bmatrix}$$

with

$$\hat{\mathbf{f}}_e(t) := \left(\int_0^{2h_e} f(t)\psi_{e,1}\mathrm{d}x, \ldots, \int_{\ell_e-2h_e}^{\ell_e} f(t)\psi_{e,N_e-1}\mathrm{d}x\right)^T,$$

$$\hat{\mathbf{f}}_{\mathcal{V}}(t) := \left(\int_{\mathscr{W}_{v_1}} f(t)\,\psi_{v_1}\mathrm{d}x, \ldots, \int_{\mathscr{W}_{vn}} f(t)\,\psi_{v_n}\mathrm{d}x\right)^T.$$

Proof The derivation of the finite element discretization in matrix form has been outlined in [2] for an elliptic PDE and the derivation of (3.7) works in the same manner. Yet, for the sake of completeness, I decided to conduct a detailed proof in section "Finite Element Semidiscretization" in Appendix. □

As in the statement of Theorem 3.6, we will omit the subscript $\boldsymbol{h}$ in the notation of the finite element solution $\mathbf{u}$ whenever the chosen discretization is clear from the context.

For later numerical computations, it will be convenient to derive a diagonal approximation of the mass matrix. This can be obtained either by lumping or by approximating the integrals in $\hat{\mathbf{M}}$ by the trapezoidal formula as proposed in [2]:

Lemma 3.7 *The computation of the integrals occurring in the mass matrix using the trapezoidal formula leads to an approximation of* $\hat{\mathbf{M}}$ *given by*

$$\bar{\mathbf{M}} := \frac{1}{2}\left(\mathrm{diag}\{(|\tilde{\mathbf{N}}\tilde{\mathbf{W}}^{-1}\tilde{\mathbf{N}}^T|)_{i,i}\}_{i=1}^{\tilde{n}}\right).$$

Proof The proof is conducted in section "Finite Element Semidiscretization" in Appendix. □

In the more general case of semilinear reaction-diffusion equations, the Galerkin approximation of Problem 2.22 on V_h reads

$$\begin{aligned}\left(u_h'(t), g_h\right)_\Gamma + \mathfrak{h}(u_h(t), g_h) &= (\mathcal{R}(u_h(t)), g_h)_\Gamma \quad \text{for all} \quad g_h \in V_h, t > 0\\ u_h(0) &= u_h^0.\end{aligned} \tag{3.8}$$

The corresponding semidiscretized IVP

$$\frac{\mathrm{d}}{\mathrm{d}t}\hat{\mathbf{M}}\mathbf{u}(t) + \hat{\mathbf{L}}\mathbf{u}(t) = \hat{\mathbf{r}}(t), \quad \mathbf{u}(0) = \mathbf{u}^0 \tag{3.9}$$

follows completely analogously to Theorem 3.6 but with $\hat{\mathbf{f}}$ replaced by

$$\hat{\mathbf{r}}(t) := \begin{bmatrix}\hat{\mathbf{r}}_{\mathcal{V}}(t)\\ \hat{\mathbf{r}}_{\mathcal{E}}(t)\end{bmatrix} \text{ where } \hat{\mathbf{r}}_{\mathcal{E}} = \begin{bmatrix}\hat{\mathbf{r}}_{e_1}(t)\\ \vdots\\ \hat{\mathbf{r}}_{e_m}(t)\end{bmatrix}^T$$

with

$$\hat{\mathbf{r}}_{\mathcal{V}}(t) := \left(\int_{\mathcal{W}_{v_1}} \mathcal{R}(u_h(t))\,\psi_{v_1}\mathrm{d}x, \ldots, \int_{\mathcal{W}_{v_n}} \mathcal{R}(u_h(t))\,\psi_{v_n}\mathrm{d}x\right)^T$$

and

$$\hat{\mathbf{r}}_e(t) := \left(\int_0^{2h_e} \mathcal{R}(u_h(t))\psi_{e,1}\mathrm{d}x, \ldots, \int_{\ell_e-2h_e}^{\ell_e} \mathcal{R}(u_h(t))\psi_{e,N_e-1}\mathrm{d}x\right)^T.$$

3.2.3 Error Analysis

An error estimate for the finite element semidiscretization can be derived using standard arguments for Galerkin finite element semidiscretizations of parabolic

PDEs (see for example [50]). The derivation in particular requires an $L^2(\Gamma)$-error estimate for the elliptic problem of finding $u \in H^1(\Gamma)$ which solves

$$\mathfrak{h}_\rho(u, g) = (f, g)_\Gamma \quad \text{for all } g \in H^1(\Gamma) \tag{3.10}$$

with $\rho = 1$, see also Problem 2.20. The desired estimate can be obtained similarly to the $H^1(\Gamma)$ estimate proven in [2] by using a duality argument. For convenience, we define $\tilde{H}^2(\Gamma) := \bigoplus_{e\in\mathcal{E}} H^2(e)$ with

$$\|u\|^2_{\tilde{H}^2(\Gamma)} := \sum_{e\in\mathcal{E}} \|u_e\|^2_{H^2(e)}$$

Note that there are no coupling conditions on the vertices in $\tilde{H}^2(\Gamma)$.

Theorem 3.8 *Let $f \in L^2(\Gamma)$ and u_h be the finite element approximation of* (3.10), *i.e.,*

$$\mathfrak{h}_1(u_h, g_h) = (f, g_h)_\Gamma \quad \textit{for all } g_h \in V_h.$$

Then, u_h satisfies

$$\|u - u_h\|_\Gamma \le \hat{c}(\Gamma)\, \hat{h}^2\, \|u\|_{\tilde{H}^2(\Gamma)}$$

where $\hat{h} := \max_{e\in\mathcal{E}} h_e$ and $\hat{c}(\Gamma)$ is a constant depending on vol_Γ.

Proof The first part of the proof is analogous to [2], Theorem 3.2 and we outline it here only for the special case $\rho = 1$. As indicated in [2], the bilinear form $\mathfrak{h}_\rho$ is coercive and continuous. In the special case $\rho = 1$, the bilinear form $\mathfrak{h}_1$ induces the $H^1(\Gamma)$-norm since $\mathfrak{h}_1(u, u) = \left(\frac{du}{dx}, \frac{du}{dx}\right)_\Gamma + (u, u)_\Gamma = \left\|\frac{du}{dx}\right\|^2_\Gamma + \|u\|^2_\Gamma = \|u\|^2_{H^1(\Gamma)}$. Due to the Galerkin orthogonality $\mathfrak{h}_1(u - u_h, g_h) = 0$ for all $g_h \in V_h$, we have

$$\|u - u_h\|_{H^1(\Gamma)} = \min\{\|u - g_h\|_{H^1(\Gamma)} : g_h \in V_h\}.$$

Together, this yields

$$\|u - u_h\|^2_{H^1(\Gamma)} \le \left\|u - u_h^I\right\|^2_{H^1(\Gamma)}$$

where $u_h^I \in V_h$ is the interpolant of u in the inner grid points and vertices. The interpolation error

$$\left\|u - u_h^I\right\|^2_{H^1(\Gamma)} = \sum_{e\in\mathcal{E}} \|u_e - (u_h^I)_e\|^2_{H^1(e)}$$

can be estimated by standard arguments since the hat functions on the edges together with the restrictions of ψ_v at the start and end vertices of the edge build a basis for a classical linear finite element approximation of $H^1(e)$. Thus, for each edge e, the central approximation theorem yields

$$\|u_e - (u_{\boldsymbol{h}}^I)_e\|_{H^1(e)} \leq c\,\ell_e h_e \|u_e\|_{H^2(e)}$$

where c is a constant independent of u and h_e. Note that since $f \in L^2(\Gamma)$, the solution satisfies $u_e \in H^2(e)$ for all $e \in \mathcal{E}$. We thus arrive at the $H^1(\Gamma)$-error estimate

$$\begin{aligned}\|u - u_{\boldsymbol{h}}\|_{H^1(\Gamma)} \leq \left(\sum_{e\in\mathcal{E}} \|u_e - (u_{\boldsymbol{h}}^I)_e\|_{H^1(e)}^2\right)^{\frac{1}{2}} &\leq c\,\mathrm{vol}_\Gamma\,\hat{h}\left(\sum_{e\in\mathcal{E}} \|u_e\|_{H^2(e)}^2\right)^{\frac{1}{2}} \\ &= c(\Gamma)\hat{h}\,\|u\|_{\tilde{H}^2(\Gamma)}\end{aligned} \tag{3.11}$$

where the constant $c(\Gamma)$ depends on the underlying metric graph, namely, its volume.

The $L^2(\Gamma)$-error estimate now follows by a classical duality argument, see for example [50], proof of Theorem 1.1. or [9], Section 0.3. The idea is to consider the solution w of the dual problem

$$\mathcal{H}w + w = u - u_{\boldsymbol{h}} \quad \text{on } \Gamma$$

subject to Neumann-Kirchhoff conditions. With the partial integration formula from Theorem 2.18, we then obtain

$$\begin{aligned}\|u - u_{\boldsymbol{h}}\|_\Gamma^2 = (u - u_{\boldsymbol{h}}, u - u_{\boldsymbol{h}})_\Gamma &= (u - u_{\boldsymbol{h}}, \mathcal{H}w + w)_\Gamma \\ &= \mathfrak{h}(u - u_{\boldsymbol{h}}, w) + (u - u_{\boldsymbol{h}}, w)_\Gamma \\ &= \mathfrak{h}_1(u - u_{\boldsymbol{h}}, w).\end{aligned}$$

Using once more the Galerkin orthogonality and the Cauchy-Schwarz inequality, we may further estimate

$$\begin{aligned}\|u - u_{\boldsymbol{h}}\|_\Gamma^2 &= \mathfrak{h}_1(u - u_{\boldsymbol{h}}, w - w_{\boldsymbol{h}}) \\ &\leq \|u - u_{\boldsymbol{h}}\|_{H^1(\Gamma)}\,\|w - w_{\boldsymbol{h}}\|_{H^1(\Gamma)}\end{aligned}$$

for $w_h \in V_h$. The $H^1(\Gamma)$-error estimate (3.11) together with the elliptic regularity inequality $\|w\|_{\tilde{H}^2(\Gamma)} \leq \hat{c}\|u - u_h\|_\Gamma$ then concludes the proof since

$$\begin{aligned}\|u - u_h\|_\Gamma &\leq \|u - u_h\|_{H^1(\Gamma)} \|w - w_h\|_{H^1(\Gamma)} / \|u - u_h\|_\Gamma \\ &\leq c_1(\Gamma)\hat{h}\, \|u\|_{\tilde{H}^2(\Gamma)} c_2(\Gamma)\hat{h}\, \|w\|_{\tilde{H}^2(\Gamma)} / \|u - u_h\|_\Gamma \\ &\leq \hat{c}(\Gamma)\hat{h}^2 \|u\|_{\tilde{H}^2(\Gamma)}.\end{aligned}$$

□

Let us now turn to the finite element semidiscretization (3.6) of the linear parabolic problem (Problem 2.16). With the previous $L^2(\Gamma)$-error estimate for the elliptic problem (3.10) and since the bilinear form fulfills *Gårding's inequality* (2.17) with $\gamma_2 = \gamma_3 = 1$, we can deduce the following standard error estimate from [50].

Theorem 3.9 *The finite element approximation of Problem 2.21 fulfills*

$$\|\boldsymbol{u}_h(t) - \boldsymbol{u}(t)\|_\Gamma \leq \left\|\boldsymbol{u}_h(0) - u^0\right\|_\Gamma + \hat{c}(\Gamma)\hat{h}^2 \left(\|u^0\|_{\tilde{H}^2(\Gamma)} + \int_0^t \|\boldsymbol{u}'\|_{\tilde{H}^2(\Gamma)} \mathrm{d}s\right)$$

for $t > 0$.

Note that in Theorem 3.9, we implicitly assume higher regularity of the solution of the continuous problem such that $\|\boldsymbol{u}'\|_{\tilde{H}^2(\Gamma)}$ is well defined. As usual, higher regularity of the solution can be deduced if the initial data are smooth and fulfill certain compatibility conditions. For the statement of Theorem 3.9, it is sufficient to require $u^0 \in \mathrm{dom}_{\mathcal{H},\mathrm{NK}}$.

Similar results can be achieved for the semilinear setting since we assumed $\mathcal{R}$ to be Lipschitz-continuous, see for example Chapter 14 in [50].

3.3 Solution of the Finite Element Semidiscretization

The objective of this section is the solution of the initial value problems

$$\frac{\mathrm{d}}{\mathrm{d}t}\hat{\mathbf{M}}\mathbf{u}(t) + \hat{\mathbf{L}}\mathbf{u}(t) = \hat{\mathbf{f}}(t), \quad \mathbf{u}(0) = \mathbf{u}^0 \tag{3.12}$$

and

$$\frac{\mathrm{d}}{\mathrm{d}t}\hat{\mathbf{M}}\mathbf{u}(t) + \hat{\mathbf{L}}\mathbf{u}(t) = \hat{\mathbf{r}}(\mathbf{u}(t)), \quad \mathbf{u}(0) = \mathbf{u}^0 \tag{3.13}$$

arising from the finite element semidiscretization of the generalized heat equation (3.6) and the reaction-diffusion equation (3.8). In the following, we only consider the more general case (3.13) and apply implicit-explicit (IMEX) schemes. In these, an implicit time stepping scheme is used for the diffusion term and an explicit scheme for the reaction term, see for example [3] for an overview of IMEX methods.

3.3.1 *Implicit-Explicit Time Stepping Schemes*

For a fixed $T > 0$, we discretize the time interval $[0, T]$ with step size Δt in time. In this context, we denote the approximate solution at time t by $\mathbf{u}^t$ and at time $t + \Delta t$ by $\mathbf{u}^{t+1}$. The simplest IMEX scheme is the forward-backward Euler discretization

$$\hat{\mathbf{M}}(\mathbf{u}^{t+1} - \mathbf{u}^t) + \Delta t\, \hat{\mathbf{L}}\mathbf{u}^{t+1} = \Delta t\, \hat{\mathbf{r}}(\mathbf{u}^t).$$

For details and higher order IMEX methods, we again refer to [3]. In the remainder of this chapter, the focus is on the efficient solution of the evolving systems of linear equations (SLE)

$$(\hat{\mathbf{M}} + \Delta t\, \hat{\mathbf{L}})\mathbf{u}^{t+1} = \hat{\mathbf{M}}\mathbf{u}^t + \Delta t\, \hat{\mathbf{r}}(\mathbf{u}^t) =: \mathbf{b} \tag{3.14}$$

with an iterative *multigrid method* (MGM). Besides the multigrid approach, we also derived a domain decomposition method based on a Schur complement identity similar to the one observed in [2]. To streamline the exposition, I decided to concentrate on the multigrid ansatz here, and, since it requires more preliminary derivations, to discuss the block decomposition approach as an alternative solver in the appendix.

Although we will outline the multigrid algorithm for the simple forward-backward Euler discretization, it can be applied straightforward to higher order IMEX methods. For example, if we apply the trapezoidal rule as implicit scheme, the evolving SLE is of the form

$$\left(\hat{\mathbf{M}} + \frac{\Delta t}{2}\hat{\mathbf{L}}\right)\mathbf{u}^{t+1} = \left(\hat{\mathbf{M}} - \frac{\Delta t}{2}\hat{\mathbf{L}}\right)\mathbf{u}^t + \Delta t\, \hat{\mathbf{r}}(\mathbf{u}^t) \tag{3.15}$$

and can be solved with a modification of the coefficient matrix and right hand side.

3.3.2 *Multigrid Solution of SLEs Arising in IMEX Schemes*

The linear system of Eqs. (3.14) can be solved by a classical multigrid generalization for metric graphs since we may interpret $\hat{\mathbf{M}} + \Delta t\, \hat{\mathbf{L}}$ as the discretization of an elliptic operator. Multigrid solutions of elliptic systems on metric graphs and their convergence are discussed in the context of a master thesis under the author's supervision[1] where elliptic equations of the type (2.14) are in focus. However, a slightly different strategy is applied here which allows to solve θ time steps simultaneously in one step of the multigrid algorithm. The presented MGM is a variation of an early approach by Hackbusch [27] who solves the discretized system

$$(\mathbf{I} + \Delta t\hat{\mathbf{L}})\mathbf{u}^{t+1} = \mathbf{u}^t,$$

also arising from an implicit Euler discretization in time.

In the following, we will only consider coarsening in space, i.e., the length of the time step Δt remains fixed. In a nutshell, the objective of a multigrid method is to solve a finite dimensional system of equations arising from a space discretization at *level* J by successively reducing it to a *coarser* discretization at level $J_0 < J$. Let us for simplicity of notation consider an equilateral metric graph for the following exposition. Then, usually and if not stated otherwise, a discretization at level J is realized with step size $\ell/2^J$ on each edge. The initial system at fine level J is given by

$$\mathbf{B}_J \mathbf{u}_J^{t+1} = \mathbf{C}_J \mathbf{u}_J^t + \mathbf{b}_J^t \tag{3.16}$$

where

$$\mathbf{B}_J := (\hat{\mathbf{M}}_J + \Delta t\, \hat{\mathbf{L}}_J), \quad \mathbf{C}_J := \hat{\mathbf{M}}_J \quad \text{and} \quad \mathbf{b}_J^t := \Delta t\, \hat{\mathbf{r}}(\mathbf{u}_J^t).$$

The vector $\mathbf{u}_J^t$ is already computed or, if $t = 0$, is the projection of the initial condition. As indicated, one step of the algorithm will not only solve for $\mathbf{u}_J^{t+1}$ but return the iterates $\mathbf{u}_J^{t+1}, \mathbf{u}_J^{t+2}, \ldots, \mathbf{u}_J^{t+\theta}$ for a preliminary chosen $\theta \in \mathbb{N}$.

One iteration (or *cycle*) of the proposed multigrid method is given in Algorithm 2 and can be described as follows.

First, a fixed number ν_1 of classical smoothing iterations is applied to the system (3.16) for the considered time points $t + \Delta t, \ldots, t + \theta\Delta t$ (*pre-smoothing*). The notation

$$\mathcal{S}^{\nu}(\mathbf{u}_J^{\tau}, \mathbf{B}_J, \mathbf{C}_J\mathbf{u}_J^{\tau-1} + \mathbf{b}_J^{\tau-1})$$

describes ν applications of the smoother to $\mathbf{u}_J^{\tau}$ for $\tau = t + 1, \ldots, t + \theta$ with coefficient matrix $\mathbf{B}_J$ and right-hand side $\mathbf{C}_J\mathbf{u}_J^{\tau-1} + \mathbf{b}_J^{\tau-1}$. Next, the *defects* (or residuals) are computed and restricted to the next coarser grid $J - 1$ by a suitable

[1] Brockmann, M.: Solving Elliptic PDEs on Metric Graphs: Finite Element Discretization, Multigrid Method and PCG Solver. Master thesis, Mathematical Institute, University of Cologne (2023).

Algorithm 2 Multigrid cycle for SLEs arising from implicit time-stepping methods

for $\tau = t+1, \dots, t+\theta$ **do** ▷ Pre-Smoothing
  $\mathbf{u}_J^\tau = \mathcal{S}^{\nu_1}(\mathbf{u}_J^\tau, \mathbf{B}_J, \mathbf{C}_J\mathbf{u}_J^{\tau-1} + \mathbf{b}_J^{\tau-1})$
end for
for $\tau = t+1, \dots, t+\theta$ **do** ▷ Defects
  $\mathbf{d}_J^\tau = \mathbf{B}_J\mathbf{u}_J^\tau - \mathbf{b}_J^\tau$
end for
for $\tau = t+1, \dots, t+\theta$ **do** ▷ Restriction
  $\mathbf{d}_{J-1}^\tau = \mathbf{R}\,\mathbf{d}_{J-1}^\tau$
end for
Solve ▷ Coarse Grid Initial Value Problem

$$\begin{aligned} \mathbf{v}_{J-1}^t &= 0 \\ \mathbf{B}_{J-1}\mathbf{v}_{J-1}^\tau &= \Delta t \mathbf{d}_{J-1}^\tau + \mathbf{C}_{J-1}\mathbf{v}_{J-1}^{\tau-1} \qquad \text{for } \tau = t+1, \dots, t+\theta \end{aligned}$$

by a direct solver (if $J = J_0$) or by η multigrid iterations
for $\tau = t+1, \dots, t+\theta$ **do** ▷ Prolongation
  $\mathbf{v}_J^\tau = \mathbf{P}\,\mathbf{v}_{J-1}^\tau$
end for
for $\tau = t+1, \dots, t+\theta$ **do** ▷ Correction
  $\mathbf{u}_J^\tau = \mathbf{u}_J^\tau - \mathbf{v}_{J-1}^\tau$
end for
for $\tau = t+1, \dots, t+\theta$ **do** ▷ Post-Smoothing
  $\mathbf{u}_J^\tau = \mathcal{S}^{\nu_2}(\mathbf{u}_J^\tau, \mathbf{B}_J, \mathbf{C}_J\mathbf{u}_J^{\tau-1} + \mathbf{b}_J^{\tau-1})$
end for

restriction operator **R** (see Sect. 3.3.3). On the coarse grid $J-1$, the residual equations are given in form of an initial value problem

$$\begin{aligned} \mathbf{v}_{J-1}^t &= 0 \\ \mathbf{B}_{J-1}\mathbf{v}_{J-1}^\tau &= \Delta t \mathbf{d}_{J-1}^\tau + \mathbf{C}_{J-1}\mathbf{v}_{J-1}^{\tau-1} \qquad \text{for } \tau = t+1, \dots, t+\theta. \end{aligned}$$

These can be solved in turn by η multigrid iterations until we arrive at the coarsest level J_0 where the initial value problem is solved by a direct solver (or an iterative method such as the conjugate gradient method). The most common choices of η are $\eta = 1$ and $\eta = 2$, referred to as V-Cycle and W-Cycle, respectively. The solutions are subsequently transferred back to the fine grid by applying a prolongation operator **P**. The iteration of the multigrid method is finally completed by the *coarse grid corrections* and, eventually, a fixed number of *post-smoothing* iterations.

The solution of the initial value problems (3.12) or (3.13) by a forward-backward Euler scheme where the arising SLEs are solved with the described multigrid approach will be referred so as implicit Euler multigrid method (IE-MGM). If the trapezoidal rule (3.15) is applied as implicit scheme, we will speak of the Crank-Nicolson multigrid method (CN-MGM).

3.3.3 Intergrid Operators

It remains to derive the restriction and prolongation operators required to transport the defect and solution from finer to coarser level and vice versa. Note that the level here refers only to the discretization in space since the time step Δt remains fixed. By levels, we thus mean extended graphs arising from the discretization with different step sizes $\boldsymbol{h}$. Again, for simplicity of exposition, suppose that we are given an equilateral graph with edge length ℓ such that the same step size h can be used at each edge. Let us in a slight abuse of notation denote by $\tilde{\Gamma}_J$ the extended graph arising from the discretization with $J = 2^{-J} - 1$ inner grid points per edge, i.e., on the coarsest level we have $\tilde{\Gamma}_0 = \Gamma$. We will moreover need the notation $\tilde{n}_J, \tilde{m}_J$ for the number of vertices and edges of $\tilde{\Gamma}_J$. Note that with our definition of levels, the finite element spaces on the single edges (3.2) are *nested*, i.e., $V_{J/2} \subset V_J$. Here, again in a slight abuse of notation, we define $V_J := V_{\ell/2^J}$.

We are interested in a prolongation operator $\mathbf{P}^J \in \mathbb{R}^{\tilde{n}_J \times \tilde{n}_{J-1}}$ interpolating the vector $\mathbf{u}^{J-1}$ from the coarser to the finer grid and a restriction operator $(\mathbf{P}^J)^T \in \mathbb{R}^{\tilde{n}_{J-1} \times \tilde{n}_J}$ for the vice versa direction. Note that we had to slightly vary the notation of the vector at level J since we will need the subscripts to refer to the entries of $\mathbf{u}^J$. The prolongation can be interpreted as a linear interpolation, compare Fig. 3.7. On the single edges, this agrees with the standard operator known in the context of classical one-dimensional discretizations on an interval, which, in a finite element

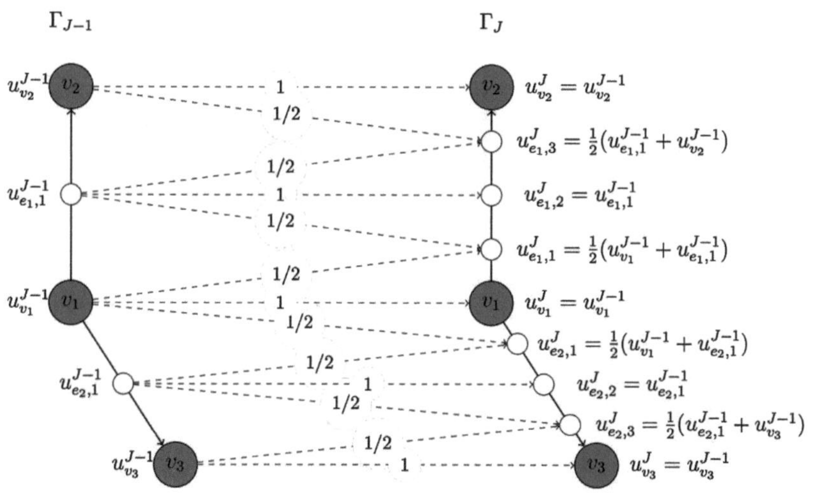

Fig. 3.7 Interpolation from coarser to finer grid

setting, can also be deduced from the refinement relation of hat basis functions. However, some special caution is required at the vertices of the graph which couple the various one-dimensional segments.

Due to the block structure of the graph, we may interpret $\mathbf{P}^J$ as a block matrix of the form

$$\mathbf{P}^J = \begin{bmatrix} \mathbf{P}^J_{\mathcal{V}\mathcal{V}} & \mathbf{P}^J_{\mathcal{V}\mathcal{E}} \\ \mathbf{P}^J_{\mathcal{E}\mathcal{V}} & \mathbf{P}^J_{\mathcal{E}\mathcal{E}} \end{bmatrix}$$

where

$$\mathbf{P}^J_{\mathcal{V}\mathcal{V}} \in \mathbb{R}^{n\times n}, \quad \mathbf{P}^J_{\mathcal{V}\mathcal{E}} \in \mathbb{R}^{n\times(\tilde{n}_{J-1}-n)}, \quad \mathbf{P}^J_{\mathcal{E}\mathcal{V}} \in \mathbb{R}^{(\tilde{n}_J-n)\times n},$$
$$\mathbf{P}^J_{\mathcal{E}\mathcal{E}} \in \mathbb{R}^{(\tilde{n}_J-n)\times(\tilde{n}_{J-1}-n)}.$$

Clearly, $\mathbf{P}^J_{\mathcal{V}\mathcal{V}} = \mathbf{I}_n$ since the index of the original vertices is maintained for each possible grid size including the original graph. Moreover, the blocks $\mathbf{P}^J_{\mathcal{E}\mathcal{E}} = \mathrm{blkDiag}\left(\{\mathbf{P}^J_e\}_{e\in\mathcal{E}}\right)$ themselves consist of m blocks associated to the original edges $e \in \mathcal{E}$. These blocks are denoted by $\mathbf{P}^J_e$ and are given by the standard operators (for the one-dimensional setting) in matrix form, i.e.,

$$\mathbf{P}^J_e = \begin{pmatrix} \frac{1}{2} & & & & & \\ 1 & & & & & \\ \frac{1}{2} & \frac{1}{2} & & & & \\ & 1 & & & & \\ & \frac{1}{2} & & & & \\ & & \cdots & & & \\ & & & \cdots & & \\ & & & & \frac{1}{2} & \\ & & & & 1 & \\ & & & & \frac{1}{2} & \frac{1}{2} \\ & & & & & 1 \\ & & & & & \frac{1}{2} \end{pmatrix}.$$

Note that $\mathbf{P}^J_e$ so far does not include the boundary of the intervals since there we have to make some adjustments according to the incidence of the first and last inner grid points to the original vertices. These will be described in $\mathbf{P}^J_{\mathcal{E}\mathcal{V}}$ as follows.

Consider a given vector $\mathbf{u}^{J-1} = \left[\mathbf{u}_{\mathcal{V}}^{J-1}, \mathbf{u}_{\mathcal{E}}^{J-1}\right]^T \in \mathbb{R}^{\tilde{n}_{J-1}}$ and its interpolation to the next finer grid $\mathbf{u}^J = \left[\mathbf{u}_{\mathcal{V}}^J, \mathbf{u}_{\mathcal{E}}^J\right]^T \in \mathbb{R}^{\tilde{n}_J}$, which both contain the values at the original vertices $\mathbf{u}_{\mathcal{V}}^{J-1}$ and $\mathbf{u}_{\mathcal{V}}^J$. Then $\mathbf{P}_{\mathcal{V}\mathcal{E}}^J$ and $\mathbf{P}_{\mathcal{E}\mathcal{V}}^J$ have to fulfill

$$\begin{bmatrix} \mathbf{P}_{\mathcal{V}\mathcal{V}}^J & \mathbf{P}_{\mathcal{V}\mathcal{E}}^J \\ \mathbf{P}_{\mathcal{E}\mathcal{V}}^J & \mathbf{P}_{\mathcal{E}\mathcal{E}}^J \end{bmatrix} \begin{pmatrix} \mathbf{u}_{\mathcal{V}}^{J-1} \\ \mathbf{u}_{\mathcal{E}}^{J-1} \end{pmatrix} = \begin{pmatrix} \mathbf{P}_{\mathcal{V}\mathcal{V}}^J \mathbf{u}_{\mathcal{V}}^{J-1} + \mathbf{P}_{\mathcal{V}\mathcal{E}}^J \mathbf{u}_{\mathcal{E}}^{J-1} \\ \mathbf{P}_{\mathcal{E}\mathcal{V}}^J \mathbf{u}_{\mathcal{V}}^{J-1} + \mathbf{P}_{\mathcal{E}\mathcal{E}}^J \mathbf{u}_{\mathcal{E}}^{J-1} \end{pmatrix} = \begin{pmatrix} \mathbf{u}_{\mathcal{V}}^J \\ \mathbf{u}_{\mathcal{E}}^J \end{pmatrix}.$$

Since the values of the original vertices are already given in the coarse graph, we have $\mathbf{u}_{\mathcal{V}}^J = \mathbf{u}_{\mathcal{V}}^{J-1}$. Consequently, with $\mathbf{P}_{\mathcal{V}\mathcal{V}}^J = \mathbf{I}_n$, we deduce $\mathbf{P}_{\mathcal{V}\mathcal{E}}^J = \mathbf{0}$. Finally, the values at the inner vertices are interpolated by

$$\mathbf{u}_{\mathcal{E}}^J = \mathbf{P}_{\mathcal{E}\mathcal{V}}^J \mathbf{u}_{\mathcal{V}}^{J-1} + \mathbf{P}_{\mathcal{E}\mathcal{E}}^J \mathbf{u}_{\mathcal{E}}^{J-1}.$$

All inner vertices that are not adjacent to an original vertex are covered by the relation $\mathbf{P}_{\mathcal{E}\mathcal{E}}^J \mathbf{u}_{\mathcal{E}}^{J-1}$ equivalent to the one-dimensional case. If we denote by $v_{e,1}^J$ and v_{e,N_e-1}^J the first and last inner vertex of $\tilde{\Gamma}_J$ on edge e, we would like the relations

$$\mathbf{u}_{e,1}^J = \frac{1}{2}\mathbf{u}_{e,0}^{J-1} + \frac{1}{2}\mathbf{u}_{e,1}^{J-1} \quad \text{and} \quad \mathbf{u}_{e,N_e-1}^J = \frac{1}{2}\mathbf{u}_{e,N_e-1}^{J-1} + \frac{1}{2}\mathbf{u}_{e,N_e}^{J-1}$$

to hold. Since $\mathbf{u}_{e,0}^{J-1}$ and $\mathbf{u}_{e,N_e}^{J-1}$ are the values at the origin and terminal vertex of e, we have to place the value $\frac{1}{2}$ in $\mathbf{P}_{\mathcal{E}\mathcal{V}}^J$ at the positions that reflect the adjacency between inner and original vertices. Using the incidence matrix

$$\tilde{\mathbf{N}}^J = \begin{bmatrix} \tilde{\mathbf{N}}_{\mathcal{V}}^J \\ \tilde{\mathbf{N}}_{\mathcal{E}}^J \end{bmatrix}$$

of $\tilde{\Gamma}_J$, we thus obtain

$$\mathbf{P}_{\mathcal{E}\mathcal{V}}^J = \frac{1}{2}\left(\left|\tilde{\mathbf{N}}_{\mathcal{E}}^J (\tilde{\mathbf{N}}_{\mathcal{V}}^J)^T\right|\right)^T.$$

Together, we derive the prolongation operator in matrix form as

$$\mathbf{P}^J = \begin{bmatrix} \mathbf{I} & \mathbf{0} \\ \frac{1}{2}\left(\left|\tilde{\mathbf{N}}_{\mathcal{E}}^J (\tilde{\mathbf{N}}_{\mathcal{V}}^J)^T\right|\right)^T & \text{blkDiag}\left(\{\mathbf{P}_e^J\}_{e\in\mathcal{E}}\right) \end{bmatrix}. \tag{3.17}$$

3.3.4 Aspects of Implementation

The MGM is an iterative method with one iteration (or cycle) consisting of the various steps outlined in Algorithm 2. In particular, an initial vector has to be chosen which in our situation is the current iterate of the implicit method $\mathbf{u}^t$ (not to be confused with the current multigrid iterate). In the elliptic situation, which will be briefly discussed in Sect. 6.4.1, a nested iteration approach is applied to acquire a suitable start vector (see for example [28], Chapter 11.5). In both cases, the systems are solved up to discretization error, where the stopping criterion is defined using the residual error.

Non-equilateral Graphs

To simplify notation, we have so far restricted to equilateral graphs in the derivation of the MGM since we can then apply a uniform discretization with $N = 2^J - 1, J \in \mathbb{N}$ inner grid points on each edge. In fact, in the non-equilateral case, it is not constructive to apply the same number of inner grid points N_e on each edge since the achieved accuracy on the edges might then differ by orders of magnitude. We rather propose to specify a maximal step size $h_{\max}$ according to the desired accuracy and then choose the number of inner grid points such that $\ell_e/N_e \leq h_{\max}$. In particular, to guarantee nested finite element spaces on each edge, we choose $N_e = 2^{J_e}$ with $J_e = \lceil \log(\ell_e/h_{\max})/\log(2) \rceil$. We can then no longer speak of a uniform level J and refer instead to the vector $(N_{e_1}, \ldots, N_{e_m})^T$ as finest level with the next coarser level $(N_{e_1}/2, \ldots, N_{e_m}/2)^T$ and so on.

In particular, we will frequently be in the situation that all inner grid points are already eliminated on several edges whereas N_e is still large on other, longer edges. We will then proceed with the restriction on the longer edges until $N_e = 1$ for all $e \in \mathcal{E}$. This situation is the coarsest possible level, equivalent to $J_0 = 0$ in the equilateral setting.

Matrix-Free Implementation

In practice, we do not need to explicitly assemble the prolongation and restriction operators but only compute their application to a vector. Given the block form (3.17), the prolongation can be performed as

$$\begin{aligned}
\mathbf{P}^J \mathbf{v}^{J-1} &= \begin{bmatrix} \mathbf{I} & \mathbf{0} \\ \frac{1}{2}\left(\left|\tilde{\mathbf{N}}_{\mathcal{E}}^J (\tilde{\mathbf{N}}_{\mathcal{V}}^J)^T\right|\right)^T & \text{blkDiag}\left(\{\mathbf{P}_e^J\}_{e\in\mathcal{E}}\right) \end{bmatrix} \begin{bmatrix} \mathbf{v}_{\mathcal{V}}^{J-1} \\ \mathbf{v}_{\mathcal{E}}^{J-1} \end{bmatrix} \\
&= \begin{bmatrix} \mathbf{v}_{\mathcal{V}}^{J-1} \\ \frac{1}{2}\left(\left|\tilde{\mathbf{N}}_{\mathcal{E}}^J (\tilde{\mathbf{N}}_{\mathcal{V}}^J)^T\right|\right)^T \mathbf{v}_{\mathcal{V}}^{J-1} + \text{blkDiag}\left(\{\mathbf{P}_e^J\}_{e\in\mathcal{E}}\right) \mathbf{v}_{\mathcal{E}}^{J-1} \end{bmatrix}.
\end{aligned}$$

The multiplications blkDiag $\left(\{\mathbf{P}_e^J\}_{e\in\mathcal{E}}\right)\mathbf{v}_{\mathcal{E}}^{J-1}$ can be performed independently for each edge and therefore reduces to m multiplications with $\mathbf{P}_e^J$. For the computation of $\frac{1}{2}\left(\left|\tilde{\mathbf{N}}_{\mathcal{E}}^J(\tilde{\mathbf{N}}_{\mathcal{V}}^J)^T\right|\right)^T \mathbf{v}_{\mathcal{V}}^{J-1}$, only $2m$ operations are necessary since for each edge only the first and last inner discretization point is adjacent to an original vertex.

In fact, the coefficient matrix $\mathbf{B}_J$ in (3.16) does not need to be explicitly assembled either. If, for example, a Jacobi smoother is applied, the smoothing step can be implemented by just computing the application of $\mathbf{B}_J$ to the current iterate. This can be realized using the expression of the mass matrix $\hat{\mathbf{M}}$ and stiffness matrix $\hat{\mathbf{L}}$ in terms of the incidence matrix of the extended graph (as derived in Theorem 3.6), i.e.,

$$\hat{\mathbf{L}} = \tilde{\mathbf{N}}\tilde{\mathbf{W}}\tilde{\mathbf{N}}^T, \quad \hat{\mathbf{M}} = \frac{1}{6}\left(|\tilde{\mathbf{N}}\tilde{\mathbf{W}}^{-1}\tilde{\mathbf{N}}^T| + \operatorname{diag}\left(\{(|\tilde{\mathbf{N}}\tilde{\mathbf{W}}^{-1}\tilde{\mathbf{N}}^T|)_{i,i}\}_{i=1}^{\tilde{n}}\right)\right).$$

The observations on the assembling of $\tilde{\mathbf{N}}$ in Sect. 3.1.3 can then be used to compute the application of the blocks

$$\tilde{\mathbf{N}}\tilde{\mathbf{W}}\tilde{\mathbf{N}}^T = \begin{bmatrix}\tilde{\mathbf{N}}_{\mathcal{V}}\\ \tilde{\mathbf{N}}_{\mathcal{E}}\end{bmatrix}\tilde{\mathbf{W}}\left[\tilde{\mathbf{N}}_{\mathcal{V}}^T\ \tilde{\mathbf{N}}_{\mathcal{E}}^T\right] = \begin{bmatrix}\tilde{\mathbf{N}}_{\mathcal{V}}\tilde{\mathbf{W}}\tilde{\mathbf{N}}_{\mathcal{V}}^T & \tilde{\mathbf{N}}_{\mathcal{V}}\tilde{\mathbf{W}}\tilde{\mathbf{N}}_{\mathcal{E}}^T\\ \tilde{\mathbf{N}}_{\mathcal{E}}\tilde{\mathbf{W}}\tilde{\mathbf{N}}_{\mathcal{V}}^T & \tilde{\mathbf{N}}_{\mathcal{E}}\tilde{\mathbf{W}}\tilde{\mathbf{N}}_{\mathcal{E}}^T\end{bmatrix}$$

(with $\tilde{\mathbf{W}}^{-1}$ respectively) on a vector.

Both a classical and a matrix free implementation have been worked out and are available in `MeGraPDE` [54]. Note that in the current implementation, a direct solver is applied to solve the systems at the coarsest level, thus the matrices have to be assembled explicitly there. If this is prohibited by the size of the initial graph, a matrix-free iterative method can be used instead.

Appendix

Finite Element Semidiscretization

Proof of Theorem 3.6 We want to approximate the solution on V_h by[2]

$$u_h(x,t) = \sum_{v\in\mathcal{V}} u_v(t)\,\psi_v(x) + \sum_{e\in\mathcal{E}}\sum_{k=1}^{N_e-1} u_{e,k}(t)\,\psi_{e,k}(x)$$

---

[2] Parts of the derivation in [2] has been supplemented in the course of a seminar work supervised by the author (Jaworski, R., Rhode, M., Schmidt, K.: Finite-Elemente-Methode fuer Quantum Graphen, seminar paper in the summer semester 2022, Mathematical Institute, University of Cologne (2022)) which served as a basis for the exposition of the proof.

with coefficients $u_{e,k}(t), u_v(t)$ that need to be determined. The objective of this proof is to derive a finite dimensional system yielding the demanded coefficients. This system, as usual, arises from the choice of test functions as basis functions of V_h, i.e., it reads

$$\frac{\mathrm{d}}{\mathrm{d}t}(u_h(t), \psi)_\Gamma + \mathfrak{h}(u_h(t), \psi) = (f(t), \psi)_\Gamma \quad \text{for all } \psi = \psi_{e,k}, \psi_v.$$

If we first test with the hat functions ψ_v defined on the vertices and then with $\psi_{e,k} \in V_{h_e}$ successively, the matrix form of this discretization has the following block structure corresponding to the original vertices and inner vertices:

$$\begin{bmatrix} \hat{\mathbf{M}}_{\mathcal{V}\mathcal{V}} & \hat{\mathbf{M}}_{\mathcal{V}\mathcal{E}} \\ \hat{\mathbf{M}}_{\mathcal{E}\mathcal{V}} & \hat{\mathbf{M}}_{\mathcal{E}\mathcal{E}} \end{bmatrix} \begin{bmatrix} \dot{\mathbf{u}}_{\mathcal{V}}(t) \\ \dot{\mathbf{u}}_{\mathcal{E}}(t) \end{bmatrix} + \begin{bmatrix} \mathbf{B}_{\mathcal{V}\mathcal{V}} & \mathbf{B}_{\mathcal{V}\mathcal{E}} \\ \mathbf{B}_{\mathcal{E}\mathcal{V}} & \mathbf{B}_{\mathcal{E}\mathcal{E}} \end{bmatrix} \begin{bmatrix} \mathbf{u}_{\mathcal{V}}(t) \\ \mathbf{u}_{\mathcal{E}}(t) \end{bmatrix} = \begin{bmatrix} \hat{\mathbf{f}}_{\mathcal{V}}(t) \\ \hat{\mathbf{f}}_{\mathcal{E}}(t) \end{bmatrix}.$$

The vector $\mathbf{u}_{\mathcal{V}}(t)$ contains the coefficients $u_v(t)$ on the original vertices and $\mathbf{u}_{\mathcal{E}}(t)$ the coefficients $u_{e,k}(t)$ on the inner vertices. We will in the following discuss how this system simplifies to the proposed matrix form and, in particular, that the stiffness matrix $\mathbf{B}$ coincides with the weighted extended graph Laplacian matrix $\hat{\mathbf{L}}$.

To start with the derivation of the stiffness matrix, we consider the bilinear form $\mathfrak{h}(\cdot,\cdot)$ and observe the following identities for our basis functions (note that for the sake of improved readability, we in the following exposition suppress the dependency of u on t):

1. Test functions on original vertices: For a fixed basis function $\psi_v \in V_h$ it holds that

$$\begin{aligned} \mathfrak{h}(u_h, \psi_v) &= \sum_{v' \in \mathcal{V}} u_v \, \mathfrak{h}(\psi_{v'}, \psi_v) + \sum_{e \in \mathcal{E}} \sum_{k=1}^{N_e - 1} u_{e,k} \, \mathfrak{h}(\psi_v, \psi_{e,k}) \\ &= u_v \int_{\mathscr{W}_v} \frac{\mathrm{d}\psi_v}{\mathrm{d}x} \frac{\mathrm{d}\psi_v}{\mathrm{d}x} \mathrm{d}x + \sum_{e \in \mathcal{E}} \sum_{k=1}^{N_e - 1} u_{e,k} \int_{\mathscr{W}_{v'} \cap e} \frac{\mathrm{d}\psi_{e,k}}{\mathrm{d}x} \frac{\mathrm{d}\psi_v}{\mathrm{d}x} \mathrm{d}x \end{aligned} \tag{3.18}$$

since $\mathscr{W}_{v'} \cap \mathscr{W}_v = \emptyset$ for $v' \neq v$.

The left part of this sum corresponds to the upper left part of the stiffness matrix $\mathbf{B}_{\mathcal{V}\mathcal{V}}$, which consequently is diagonal with entries given by

$$(\mathbf{B}_{\mathcal{V}\mathcal{V}})_{v,v} = \int_{\mathscr{W}_v} \frac{\mathrm{d}\psi_v}{\mathrm{d}x} \frac{\mathrm{d}\psi_v}{\mathrm{d}x} \mathrm{d}x.$$

Since

$$\begin{aligned}
\int_{\mathscr{W}_v} \frac{\mathrm{d}\psi_v}{\mathrm{d}x}\frac{\mathrm{d}\psi_v}{\mathrm{d}x}\mathrm{d}x &= \sum_{e\in\mathcal{E}_v^{\mathrm{out}}} \int_{\mathscr{W}_v\cap e} \frac{\mathrm{d}\psi_v}{\mathrm{d}x}\frac{\mathrm{d}\psi_v}{\mathrm{d}x}\mathrm{d}x + \sum_{e\in\mathcal{E}_v^{\mathrm{in}}} \int_{\mathscr{W}_v\cap e} \frac{\mathrm{d}\psi_v}{\mathrm{d}x}\frac{\mathrm{d}\psi_v}{\mathrm{d}x}\mathrm{d}x \\
&= \sum_{e\in\mathcal{E}_v^{\mathrm{out}}} \int_0^{h_e} \left(-\frac{1}{h_e}\right)^2 \mathrm{d}x + \sum_{e\in\mathcal{E}_v^{\mathrm{in}}} \int_{\ell_e-h_e}^{\ell_e} \left(\frac{1}{h_e}\right)^2 \mathrm{d}x \\
&= \sum_{e\in\mathcal{E}_v^{\mathrm{out}}} \frac{1}{h_e} + \sum_{e\in\mathcal{E}_v^{\mathrm{in}}} \frac{1}{h_e} = \sum_{e\in\mathcal{E}_v} \frac{1}{h_e},
\end{aligned}$$

these exactly agree with the entries of $\hat{\mathbf{L}}_{\mathcal{V}\mathcal{V}}$.

Consider now the right part of (3.18) corresponding to the upper right part of the stiffness matrix. The supports of $\psi_{e,k}$ and ψ_v only overlap if $e \in \mathcal{E}_v$ and k is the first or last inner grid point on e. More precisely, it holds that

$$\operatorname{supp}(\psi_{e,k} \cap \psi_v) = \begin{cases} [0, h_e] & \text{if } e \in \mathcal{E}_v^{\mathrm{out}} \text{ and } k = 1, \\ [\ell_e - h_e, \ell_e] & \text{if } e \in \mathcal{E}_v^{\mathrm{in}} \text{ and } k = N_e - 1, \\ 0 & \text{otherwise.} \end{cases} \tag{3.19}$$

We deduce that

$$\begin{aligned}
&\sum_{e\in\mathcal{E}} \sum_{k=1}^{N_e-1} u_{e,k} \int_{\mathscr{W}_v\cap e} \frac{\mathrm{d}\psi_{e,k}}{\mathrm{d}x}\frac{\mathrm{d}\psi_v}{\mathrm{d}x}\mathrm{d}x \\
&= \sum_{e\in\mathcal{E}_v^{\mathrm{out}}} u_{e,1} \int_0^{h_e} \frac{\mathrm{d}\psi_{e,1}}{\mathrm{d}x}\frac{\mathrm{d}\psi_v}{\mathrm{d}x}\mathrm{d}x + \sum_{e\in\mathcal{E}_v^{\mathrm{in}}} u_{e,N_e-1} \int_{\ell_e-h_e}^{\ell_e} \frac{\mathrm{d}\psi_{e,N_e-1}}{\mathrm{d}x}\frac{\mathrm{d}\psi_v}{\mathrm{d}x}\mathrm{d}x \\
&= \sum_{e\in\mathcal{E}_v^{\mathrm{out}}} u_{e,1} \int_0^{h_e} \frac{1}{h_e}\left(-\frac{1}{h_e}\right)\mathrm{d}x + \sum_{e\in\mathcal{E}_v^{\mathrm{in}}} u_{e,N_e-1} \int_{\ell_e-h_e}^{\ell_e} \left(-\frac{1}{h_e}\right)\frac{1}{h_e}\mathrm{d}x \\
&= \sum_{e\in\mathcal{E}_v^{\mathrm{out}}} u_{e,1}\left(-\frac{1}{h_e}\right) + \sum_{e\in\mathcal{E}_v^{\mathrm{in}}} u_{e,N_e-1}\left(-\frac{1}{h_e}\right),
\end{aligned}$$

i.e., $(\mathbf{B}_{\mathcal{V}\mathcal{E}})_{v,\tilde{v}} = -\frac{1}{h_e}$ if the original vertex v is connected to the inner grid point $\tilde{v}$ and zero otherwise. This again coincides with the weighted extended graph Laplacian matrix $\hat{\mathbf{L}}_{\mathcal{V}\mathcal{E}}$.

2. Test functions on inner vertices: For a fixed basis function $(\psi_{e,k}) \in \mathcal{V}_{h_e}$ on $e \in \mathcal{E}$ it holds that

$$\begin{aligned}\mathfrak{h}(\boldsymbol{u_h}, \psi_{e,k}) &= \sum_{e' \in \mathcal{E}} \sum_{j=1}^{N_e-1} u_{e,j}\, \mathfrak{h}(\psi_{e',j}, \psi_{e,k}) + \sum_{v \in \mathcal{V}} u_v\, \mathfrak{h}(\psi_v, \psi_{e,k}) \\ &= \sum_{j=1}^{N_e-1} u_{e,j} \int_e \frac{\mathrm{d}\psi_{e,j}}{\mathrm{d}x} \frac{\mathrm{d}\psi_{e,k}}{\mathrm{d}x}\, \mathrm{d}x + \sum_{v \in \mathcal{V}} u_v \int_{\mathcal{W}_v \cap e} \frac{\mathrm{d}\psi_v}{\mathrm{d}x} \frac{\mathrm{d}\psi_{e,k}}{\mathrm{d}x}\, \mathrm{d}x\end{aligned}$$

since $\displaystyle\int_{e'} \frac{\mathrm{d}\psi_{e',j}}{\mathrm{d}x} \frac{\mathrm{d}\psi_{e,k}}{\mathrm{d}x} = 0$ for $e' \neq e$.

The second part of the sum corresponds to the upper left part of the stiffness matrix and, similar to $\mathbf{B}_{\mathcal{V}\mathcal{E}}$, we obtain

$$\int_{\mathcal{W}_v \cap e} \frac{\mathrm{d}\psi_v}{\mathrm{d}x} \frac{\mathrm{d}\psi_{e,k}}{\mathrm{d}x}\, \mathrm{d}x = -\frac{1}{h_e}.$$

This means that $(\mathbf{B}_{\mathcal{E}\mathcal{V}})_{\tilde{v},v}$ is equal to $-\frac{1}{h_e}$ if the inner grid point $\tilde{v}$ is connected to the original vertex v and zero otherwise, i.e., we have $\mathbf{B}_{\mathcal{E}\mathcal{V}} = \mathbf{B}_{\mathcal{V}\mathcal{E}}^T = \hat{\mathbf{L}}_{\mathcal{V}\mathcal{E}}^T$.

Finally, the first part of the sum describes the entries of the lower right part of the stiffness matrix. Clearly, the supports only overlap if $e = e'$ and as

$$\operatorname{supp}(\psi_{e,k} \cap \psi_{e,j}) = \begin{cases} [x_{e,j-1}, x_{e,j}] & \text{if } k = j-1, \\ [x_{e,j-1}, x_{e,j+1}] & \text{if } k = j, \\ [x_{e,j}, x_{e,j+1}] & \text{if } k = j+1, \\ 0 & \text{otherwise,} \end{cases} \tag{3.20}$$

we for each edge distinguish the three cases:

1. If $k = j$ it holds that:

$$\begin{aligned}\int_e \frac{\mathrm{d}\psi_{e,k}}{\mathrm{d}x} \frac{\mathrm{d}\psi_{e,j}}{\mathrm{d}x} \mathrm{d}x &= \int_{x_{e,j-1}}^{x_{e,j}} \left(\frac{\mathrm{d}\psi_{e,j}}{\mathrm{d}x}\right)^2 \mathrm{d}x + \int_{x_{e,j}}^{x_{e,j+1}} \left(\frac{\mathrm{d}\psi_{e,j}}{\mathrm{d}x}\right)^2 \mathrm{d}x \\ &= \int_{x_{e,j-1}}^{x_{e,j}} \left(\frac{1}{h_e}\right)^2 \mathrm{d}x + \int_{x_{e,j}}^{x_{e,j+1}} \left(-\frac{1}{h_e}\right)^2 \mathrm{d}x \\ &= \frac{2}{h_e}.\end{aligned}$$

2. If $k = j-1$ it holds that:

$$\int_e \frac{\mathrm{d}\psi_{e,j-1}}{\mathrm{d}x} \frac{\mathrm{d}\psi_{e,j}}{\mathrm{d}x} \mathrm{d}x = \int_{x_{e,j-1}}^{x_{e,j}} -\frac{1}{h_e} \frac{1}{h_e} \mathrm{d}x = -\frac{1}{h_e}.$$

3. The case $k = j + 1$ follows equivalently to the second case.

Consequently, $\mathbf{B}_{\mathcal{E}\mathcal{E}} = \text{blkDiag}(\{\mathbf{B}_e\}_{e\in\mathcal{E}})$ is a block diagonal matrix with tridiagonal blocks

$$\mathbf{B}_e = \frac{1}{h_e}\begin{pmatrix} 2 & -1 & & & \\ -1 & 2 & -1 & & \\ & \ddots & \ddots & \ddots & \\ & & -1 & 2 & -1 \\ & & & -1 & 2 \end{pmatrix}$$

what exactly defines $\hat{\mathbf{L}}_{\mathcal{E}\mathcal{E}}$.

Thus, the finite element approximation of the bilinear form can be represented with the weighted graph Laplacian matrix of the extended graph.

To show the identity of the mass matrix, we first observe that $|\tilde{\mathbf{N}}\tilde{\mathbf{W}}^{-1}\tilde{\mathbf{N}}^T|$ coincides with the absolute graph Laplacian matrix of $\tilde{\Gamma}$ with inverse edge weights, i.e. with weights $\ell_{\tilde{e}}$. Moreover, $\text{diag}\left(\{(|\tilde{\mathbf{N}}\tilde{\mathbf{W}}^{-1}\tilde{\mathbf{N}}^T|)_{i,i}\}_{i=1}^{\tilde{n}}\right)$ describes the diagonal of this matrix. This means we have to show that the blocks of $\hat{\mathbf{M}}$ are of the following form:

1. $\hat{\mathbf{M}}_{\mathcal{V}\mathcal{V}} = \frac{1}{6}\text{diag}\left(\left\{\sum_{e\in\mathcal{E}_v} 2h_e\right\}_{v\in\mathcal{V}}\right)$
2. $\hat{\mathbf{M}}_{\mathcal{E}\mathcal{E}}$ is a block diagonal matrix $\hat{\mathbf{M}}_{\mathcal{E}\mathcal{E}} = \text{blkDiag}(\{\hat{\mathbf{M}}_e\}_{e\in\mathcal{E}})$ with tridiagonal blocks

$$\hat{\mathbf{M}}_e = \frac{h_e}{6}\begin{pmatrix} 4 & 1 & & & \\ 1 & 4 & 1 & & \\ & \ddots & \ddots & \ddots & \\ & & 1 & 4 & 1 \\ & & & 1 & 4 \end{pmatrix}$$

3. $\hat{\mathbf{M}}_{\mathcal{V}\mathcal{E}}$ has entries at the positions where original vertices v and inner vertices $\tilde{v} \in e \in \mathcal{E}_v$ are connected, in particular these entries are given by $\frac{h_e}{6}$.
4. $\hat{\mathbf{M}}_{\mathcal{E}\mathcal{V}} = \hat{\mathbf{M}}_{\mathcal{V}\mathcal{E}}^T$.

To derive the proposed blocks, we consider

$$\begin{aligned}\frac{\mathrm{d}}{\mathrm{d}t}(\boldsymbol{u_h}, \psi)_\Gamma &= \frac{\mathrm{d}}{\mathrm{d}t}\left(\sum_{v\in\mathcal{V}} u_v(t)\,\psi_v(x) + \sum_{e\in\mathcal{E}}\sum_{k=1}^{N_e-1} u_{e,k}(t)\,\psi_{e,k}(x)\,,\,\psi(x)\right)_\Gamma \\ &= \int_\Gamma\left(\sum_{v\in\mathcal{V}} \frac{\partial}{\partial t}u_v(t)\,\psi_v(x)\psi(x) + \sum_{e\in\mathcal{E}}\sum_{k=1}^{N_e-1} u_{e,k}(t)\psi_{e,k}(x)\psi(x)\right)\mathrm{d}x\end{aligned}$$

for all $\psi \in V_h$. We can then again simplify the blocks of $\hat{\mathbf{M}}$ by the following distinction of the test functions:

1. Test functions on original vertices: For a fixed basis function $\psi_v \in V_h$ it holds that

$$\int_\Gamma \left(\sum_{v' \in \mathcal{V}} \frac{\partial}{\partial t} u_{v'}(t)\, \psi_{v'}(x) \psi_v(x) + \sum_{e \in \mathcal{E}} \sum_{k=1}^{N_e - 1} u_{e,k}(t) \psi_{e,k}(x) \psi_v(x) \right) \mathrm{d}x$$
$$= \frac{\mathrm{d}}{\mathrm{d}t} u_v(t) \int_\Gamma \psi_v(x) \psi_v(x) \mathrm{d}x + \int_\Gamma \sum_{e \in \mathcal{E}} \sum_{k=1}^{N_e - 1} u_{e,k}(t) \psi_{e,k}(x) \psi_v(x) \mathrm{d}x$$

with the arguments on the supports of $\psi_v(x)$ as above. Thus, $\hat{\mathbf{M}}_{\mathcal{V}\mathcal{V}}$ is diagonal with

$$(\hat{\mathbf{M}}_{\mathcal{V}\mathcal{V}})_{v,v} = \int_\Gamma \psi_v(x) \psi_v(x) \mathrm{d}x = \int_{\mathscr{W}_v} \psi_v(x) \psi_v(x) \mathrm{d}x. \tag{3.21}$$

These integrals may be calculated exactly as

$$\int_{\mathscr{W}_v} \psi_v(x) \psi_v(x) \mathrm{d}x = \sum_{e \in \mathcal{E}_v^{\mathrm{out}}} \int_{\mathscr{W}_v \cap e} \psi_v(x) \psi_v(x) \mathrm{d}x + \sum_{e \in \mathcal{E}_v^{\mathrm{in}}} \int_{\mathscr{W}_v \cap e} \psi_v(x) \psi_v(x) \mathrm{d}x$$
$$= \sum_{e \in \mathcal{E}_v^{\mathrm{out}}} \int_0^{h_e} \left(1 - \frac{x}{h_e}\right)^2 \mathrm{d}x + \sum_{e \in \mathcal{E}_v^{\mathrm{in}}} \int_{\ell_e - h_e}^{\ell_e} \left(1 - \frac{\ell_e - x}{h_e}\right)^2 \mathrm{d}x$$
$$= \sum_{e \in \mathcal{E}_v^{\mathrm{out}}} \frac{h_e}{3} + \sum_{e \in \mathcal{E}_v^{\mathrm{in}}} \frac{h_e}{3} = \sum_{e \in \mathcal{E}_v} \frac{h_e}{3}.$$

Moreover, the right part of the sum simplifies to

$$\int_\Gamma \left(\sum_{e \in \mathcal{E}} \sum_{k=1}^{N_e - 1} u_{e,k}(t) \psi_{e,k}(x) \psi_v(x) \right) \mathrm{d}x = \sum_{e \in \mathcal{E}_v} \sum_{k=1}^{N_e - 1} u_{e,k}(t) \int_{\mathscr{W}_v} \psi_{e,k}(x) \psi_v(x) \mathrm{d}x \tag{3.22}$$

and, by (3.19), further to

$$= \sum_{e \in \mathcal{E}_v} \sum_{k=1}^{N_e - 1} u_{e,k}(t) \int_{\mathscr{W}_v} \psi_{e,k}(x) \psi_v(x) \mathrm{d}x$$
$$= \sum_{e \in \mathcal{E}_v^{\mathrm{out}}} u_{e,1}(t) \int_0^{h_e} \psi_{e,1}(x) \psi_v(x) \mathrm{d}x$$
$$+ \sum_{e \in \mathcal{E}_v^{\mathrm{in}}} u_{e,N_e - 1}(t) \int_{\ell_e - h_e}^{\ell_e} \psi_{e,N_e - 1}(x) \psi_v(x) \mathrm{d}x$$

with

$$\int_0^{h_e} \psi_{e,1}(x)\psi_v(x)\mathrm{d}x = \int_0^{h_e} \left(1 - \frac{x_{e,1} - x}{h_e}\right)\left(1 - \frac{x}{h_e}\right)\mathrm{d}x = \frac{h_e}{6}$$

and

$$\int_{\ell_e - h_e}^{\ell_e} \psi_{e,N_e-1}(x)\psi_v(x)\mathrm{d}x = \int_{\ell_e - h_e}^{\ell_e} \left(1 + \frac{x_{e,N_e-1} - x}{h_e}\right)\left(1 - \frac{\ell_e - x}{h_e}\right)\mathrm{d}x = \frac{h_e}{6}.$$

Consequently, the entries of $\hat{\mathbf{M}}_{\mathcal{V}\mathcal{E}}$ are given by $\frac{h_e}{6}$ whenever an original vertex is connected to an inner grid point.

2. Test functions on inner discretization points: For a fixed basis function $\psi_{e,k} \in V_h$ it holds that

$$\begin{aligned}
&\int_\Gamma \left(\sum_{v \in \mathcal{V}} \frac{\mathrm{d}}{\mathrm{d}t} u_v(t)\, \psi_v(x)\psi_{e,k}(x) + \sum_{e' \in \mathcal{E}} \sum_{j=1}^{N_e - 1} u_{e',j}(t)\psi_{e',j}(x)\psi_{e,k}(x) \right) \mathrm{d}x \\
&= \int_\Gamma \left(\sum_{v \in \mathcal{V}} \frac{\mathrm{d}}{\mathrm{d}t} u_v(t)\, \psi_v(x)\psi_{e,k}(x) + \sum_{j=1}^{N_e - 1} u_{e,j}(t)\psi_{e,j}(x)\psi_{e,k}(x) \right) \mathrm{d}x,
\end{aligned} \tag{3.23}$$

i.e., to determine the entries of $\hat{\mathbf{M}}_{\mathcal{E}\mathcal{E}}$ we consider

$$\int_\Gamma \sum_{j=1}^{N_e - 1} u_{e,j}(t)\psi_{e,j}(x)\psi_{e,k}(x)\mathrm{d}x = \sum_{j=1}^{N_e - 1} u_{e,j} \int_e \psi_{e,j}(x)\psi_{e,k}(x)\mathrm{d}x \tag{3.24}$$

and using (3.20) calculate the integrals for the three occurring cases:

1. If $j = k$, we obtain

$$\begin{aligned}
\int_e (\psi_{e,k}(x))^2 \mathrm{d}x &= \int_{x_{e,k-1}}^{x_{e,k}} \left(1 - \frac{x_{e,k} - x}{h_e}\right)^2 \mathrm{d}x + \int_{x_{e,k}}^{x_{e,k+1}} \left(1 + \frac{x_{e,k} - x}{h_e}\right)^2 \mathrm{d}x \\
&= 2\frac{h_e}{3},
\end{aligned}$$

2. if $j = k - 1$, we have

$$\int_e \psi_{e,k-1}(x)\psi_{e,k}(x)\mathrm{d}x = \int_{x_{e,k-1}}^{x_{e,k}} \left(1 + \frac{x_{e,k-1} - x}{h_e}\right)\left(1 - \frac{x_{e,k} - x}{h_e}\right)\mathrm{d}x = \frac{h_e}{6}$$

3. and the case $j = k + 1$ it follows similar to the second case.

The left part of the sum in (3.23) corresponds to $\hat{\mathbf{M}}_{\mathcal{E}\mathcal{V}}$ and follows equivalent to the derivation of $\hat{\mathbf{M}}_{\mathcal{V}\mathcal{E}}$ since $\hat{\mathbf{M}}_{\mathcal{E}\mathcal{V}} = \hat{\mathbf{M}}_{\mathcal{V}\mathcal{E}}^T$.

Finally, the characterization of the right hand side immediately follows taking into account the supports of the test functions. □

Proof of Lemma 3.7 We go through the blocks of the mass matrix

$$\hat{\mathbf{M}} = \begin{bmatrix} \hat{\mathbf{M}}_{\mathcal{V}\mathcal{V}} & \hat{\mathbf{M}}_{\mathcal{V}\mathcal{E}} \\ \hat{\mathbf{M}}_{\mathcal{E}\mathcal{V}} & \hat{\mathbf{M}}_{\mathcal{E}\mathcal{E}} \end{bmatrix}$$

sequentially and approximate the integrals using the trapezoidal rule.

1. The integrals in the diagonal matrix $\hat{\mathbf{M}}_{\mathcal{V}\mathcal{V}}$ are given by

$$\begin{aligned} \int_{\mathscr{W}_v} \psi_v(x)\psi_v(x)\mathrm{d}x &\approx \sum_{e\in\mathcal{E}_v^{\mathrm{out}}} \int_0^{h_e} \left(1 - \frac{x}{h_e}\right)^2 \mathrm{d}x + \sum_{e\in\mathcal{E}_v^{\mathrm{in}}} \int_{\ell_e - h_e}^{\ell_e} \left(1 - \frac{\ell_e - x}{h_e}\right)^2 \mathrm{d}x \\ &= \sum_{e\in\mathcal{E}_v^{\mathrm{out}}} \frac{h_e}{2} + \sum_{e\in\mathcal{E}_v^{\mathrm{in}}} \frac{h_e}{2} = \frac{1}{2}\sum_{e\in\mathcal{E}} h_e. \end{aligned}$$

2. For the integrals of the block diagonal matrix $\hat{\mathbf{M}}_{\mathcal{E}\mathcal{E}} = \mathrm{blkDiag}(\{\hat{\mathbf{M}}_e\}_{e\in\mathcal{E}})$ we have

$$(\hat{\mathbf{M}}_e)_{k,j} = \int_e \psi_{e,k}(x)\psi_{e,j}(x)\mathrm{d}x$$

with three cases:

a. If $j = k$:

$$\begin{aligned} \int_e (\psi_{e,k}(x))^2 \mathrm{d}x &= \int_{x_{e,k-1}}^{x_{e,k}} \left(1 - \frac{x_{e,k} - x}{h_e}\right)^2 \mathrm{d}x + \int_{x_{e,k}}^{x_{e,k+1}} \left(1 + \frac{x_{e,k} - x}{h_e}\right)^2 \mathrm{d}x \\ &\approx \frac{h_e}{2} + \frac{h_e}{2} = h_e, \end{aligned}$$

b. if $j = k - 1$:

$$\int_e \psi_{e,k}(x)\psi_{e,k-1}(x)\mathrm{d}x = \int_{x_{e,k-1}}^{x_{e,k}} \left(1 + \frac{x_{e,k-1} - x}{h_e}\right)\left(1 - \frac{x_{e,k} - x}{h_e}\right)\mathrm{d}x \approx 0$$

c. and $j = k + 1$ equivalent to 2.

3. The integrals in $\hat{\mathbf{M}}_{\mathcal{V}\mathcal{E}}$ and $\hat{\mathbf{M}}_{\mathcal{E}\mathcal{V}}$ are

$$\int_0^{h_e} \psi_{e,1}(x)\psi_v(x)\mathrm{d}x = \int_0^{h_e} \left(1 - \frac{x_{e,1} - x}{h_e}\right)\left(1 - \frac{x}{h_e}\right)\mathrm{d}x$$

and

$$\int_{\ell_e - h_e}^{\ell_e} \psi_{e,N_e-1}(x)\psi_v(x)\mathrm{d}x = \int_{\ell_e - h_e}^{\ell_e} \left(1 + \frac{x_{e,N_e-1} - x}{h_e}\right)\left(1 - \frac{\ell_e - x}{h_e}\right)\mathrm{d}x,$$

which, approximated by the trapezoidal rule, are both equal to zero. Consequently, $\hat{\mathbf{M}}_{\mathcal{V}\mathcal{E}} = \hat{\mathbf{M}}_{\mathcal{E}\mathcal{V}}^T = \mathbf{0}$.

Thus, $\bar{\mathbf{M}}$ is a diagonal matrix and since the inner vertices have degree two, it is exactly half the degree matrix of the extended graph with inverse edge weights h_e.

□

Schur Complement

The extended graph delivers a natural domain decomposition in original vertices and inner grid points. Motivated by this fact, in [2], the authors discuss a block decomposition approach for the solution of systems involving $\tilde{\mathbf{L}}$. For equilateral graphs, they prove the identity of the Schur complement of $\tilde{\mathbf{L}}$ and the Laplacian matrix $\mathbf{L}$ of the original, undiscretized graph ([2], Theorem 4.3.). With the help of the introduced notation for weighted graph matrices, we will generalize this identity for arbitrary, non-equilateral graphs.

Theorem 3.10 *Let Γ be a metric graph and $\tilde{\Gamma}$ be the extended graph arising from the discretization with step sizes $h_e \in \boldsymbol{h}$ on the edges. We assign to the edges of both graphs a weight according to the reciprocal of their length, i.e., $w_e = 1/\ell_e$ for $e \in \mathcal{E}$ and $w_{\tilde{e}} = 1/h_e$ for $\tilde{e} \in \tilde{\mathcal{E}}_e$ and all $e \in \mathcal{E}$. Then, the Schur complement of $\tilde{\mathbf{L}}$ in $\hat{\mathbf{L}}_{\mathcal{E}\mathcal{E}}$ is given by*

$$\hat{\mathbf{S}} := \hat{\mathbf{L}}/\hat{\mathbf{L}}_{\mathcal{E}\mathcal{E}} = \hat{\mathbf{L}}_{\mathcal{V}\mathcal{V}} - \hat{\mathbf{L}}_{\mathcal{V}\mathcal{E}}\hat{\mathbf{L}}_{\mathcal{E}\mathcal{E}}^{-1}\hat{\mathbf{L}}_{\mathcal{V}\mathcal{E}}^T = \hat{\mathbf{L}}_\Gamma$$

where $\hat{\mathbf{L}}_\Gamma$ is the weighted Laplacian matrix of the original graph Γ.

Proof The proof works more or less equivalent to the equilateral case, which is given in [2], proof of Theorem 4.3. However, some of the generalizations and simplifications eligible in the equilateral case do not apply here since the blocks of $\hat{\mathbf{L}}_{\mathcal{E}\mathcal{E}}$ no longer have the same size and weights. We therefore provide a detailed proof of the assertion:

The extended graph Laplacian matrix can be computed as $\hat{\mathbf{L}} = \tilde{\mathbf{N}}\tilde{\mathbf{W}}\tilde{\mathbf{N}}^T$ with $\tilde{\mathbf{W}} = \operatorname{diag}((w_{\tilde{e}})_{\tilde{e}\in\tilde{\mathcal{E}}})$. Consequently, the blocks of $\hat{\mathbf{L}}$ are given by

$$\begin{bmatrix} \hat{\mathbf{L}}_{\mathcal{V}\mathcal{V}} & \hat{\mathbf{L}}_{\mathcal{V}\mathcal{E}} \\ \hat{\mathbf{L}}_{\mathcal{E}\mathcal{V}} & \hat{\mathbf{L}}_{\mathcal{E}\mathcal{E}} \end{bmatrix} = \begin{bmatrix} \tilde{\mathbf{N}}_{\mathcal{V}} \\ \tilde{\mathbf{N}}_{\mathcal{E}} \end{bmatrix} \tilde{\mathbf{W}} \left[\tilde{\mathbf{N}}_{\mathcal{V}}^T \; \tilde{\mathbf{N}}_{\mathcal{E}}^T\right] = \begin{bmatrix} \tilde{\mathbf{N}}_{\mathcal{V}}\tilde{\mathbf{W}}\tilde{\mathbf{N}}_{\mathcal{V}}^T & \tilde{\mathbf{N}}_{\mathcal{V}}\tilde{\mathbf{W}}\tilde{\mathbf{N}}_{\mathcal{E}}^T \\ \tilde{\mathbf{N}}_{\mathcal{E}}\tilde{\mathbf{W}}\tilde{\mathbf{N}}_{\mathcal{V}}^T & \tilde{\mathbf{N}}_{\mathcal{E}}\tilde{\mathbf{W}}\tilde{\mathbf{N}}_{\mathcal{E}}^T \end{bmatrix}. \tag{3.25}$$

Consider the Schur complement

$$\hat{\mathbf{S}} := \hat{\mathbf{L}}/\hat{\mathbf{L}}_{\mathcal{E}\mathcal{E}} = \hat{\mathbf{L}}_{\mathcal{V}\mathcal{V}} - \hat{\mathbf{L}}_{\mathcal{V}\mathcal{E}}\hat{\mathbf{L}}_{\mathcal{E}\mathcal{E}}^{-1}\hat{\mathbf{L}}_{\mathcal{E}\mathcal{V}}$$

which is well defined since $\hat{\mathbf{L}}_{\mathcal{E}\mathcal{E}}$ is not singular. With (3.25), the second summand can be rewritten to

$$\hat{\mathbf{L}}_{\mathcal{V}\mathcal{E}}\hat{\mathbf{L}}_{\mathcal{E}\mathcal{E}}^{-1}\hat{\mathbf{L}}_{\mathcal{E}\mathcal{V}} = \tilde{\mathbf{N}}_{\mathcal{V}} \left(\tilde{\mathbf{W}}\tilde{\mathbf{N}}_{\mathcal{E}}^T \left(\tilde{\mathbf{N}}_{\mathcal{E}}\tilde{\mathbf{W}}\tilde{\mathbf{N}}_{\mathcal{E}}^T\right)^{-1} \tilde{\mathbf{N}}_{\mathcal{E}}\tilde{\mathbf{W}} \right) \tilde{\mathbf{N}}_{\mathcal{V}}^T. \tag{3.26}$$

Note that

$$\tilde{\mathbf{W}} = \operatorname{blkDiag}\left(\left\{\frac{1}{h_e}\mathbf{I}_{N_e}\right\}_{e\in\mathcal{E}}\right) \quad \text{and} \quad \tilde{\mathbf{N}}_{\mathcal{E}} = \operatorname{blkDiag}\left(\{\tilde{\mathbf{N}}_e\}_{e\in\mathcal{E}}\right)$$

with $\tilde{\mathbf{N}}_e \in \mathbb{R}^{(N_e-1)\times N_e}$ and, consequently, $\tilde{\mathbf{N}}_{\mathcal{E}}\tilde{\mathbf{W}}\tilde{\mathbf{N}}_{\mathcal{E}}^T$ is a blockdiagonal matrix with its inverse given by

$$\left(\tilde{\mathbf{N}}_{\mathcal{E}}\tilde{\mathbf{W}}\tilde{\mathbf{N}}_{\mathcal{E}}^T\right)^{-1} = \operatorname{blkDiag}\left(\left\{h_e\left(\tilde{\mathbf{N}}_e\tilde{\mathbf{N}}_e^T\right)^{-1}\right\}_{e\in\mathcal{E}}\right).$$

Moreover, $\tilde{\mathbf{W}}\tilde{\mathbf{N}}_{\mathcal{E}}^T$ is a blockdiagonal matrix of the form $\tilde{\mathbf{W}}\tilde{\mathbf{N}}_{\mathcal{E}}^T = \operatorname{blkDiag}\left(\{\frac{1}{h_e}\tilde{\mathbf{N}}_e^T\}_{e\in\mathcal{E}}\right)$. Together, the expression in the middle of (3.26) is given by

$$\tilde{\mathbf{W}}\tilde{\mathbf{N}}_{\mathcal{E}}^T \left(\tilde{\mathbf{N}}_{\mathcal{E}}\tilde{\mathbf{W}}\tilde{\mathbf{N}}_{\mathcal{E}}^T\right)^{-1} \tilde{\mathbf{N}}_{\mathcal{E}}\tilde{\mathbf{W}} = \operatorname{blkDiag}\left(\left\{\frac{1}{h_e}\tilde{\mathbf{N}}_e^T\left(\tilde{\mathbf{N}}_e\tilde{\mathbf{N}}_e^T\right)^{-1}\tilde{\mathbf{N}}_e\right\}_{e\in\mathcal{E}}\right).$$

If we denote by $\mathbf{1}_{N_e} \in \mathbb{R}^{N_e}$ the vector of all ones, each of these blocks can be expressed as

$$\frac{1}{h_e}\tilde{\mathbf{N}}_e^T\left(\tilde{\mathbf{N}}_e\tilde{\mathbf{N}}_e^T\right)^{-1}\tilde{\mathbf{N}}_e = \frac{1}{h_e}\frac{1}{N_e}\left(N_e\mathbf{I}_{N_e} - \mathbf{1}_{N_e}\mathbf{1}_{N_e}^T\right) =: \mathbf{T}_e,$$

see Lemma 3.11.

Remember now from the previous section that

$$\tilde{\mathbf{N}}_{\mathcal{V}} = \left[\tilde{\mathbf{N}}_{\mathcal{V}_{e_1}}, \ldots, \tilde{\mathbf{N}}_{\mathcal{V}_{em}}\right]$$

with

$$\tilde{\mathbf{N}}_{\mathcal{V}_e} = \mathbf{N}_e^{\text{out}} \otimes (\mathbf{e}_1^{N_e})^T + \mathbf{N}_e^{\text{in}} \otimes (\mathbf{e}_{N_e}^{N_e})^T.$$

We deduce that (3.26) can be expressed as

$$\begin{aligned}
&\tilde{\mathbf{N}}_{\mathcal{V}} \left(\tilde{\mathbf{W}} \tilde{\mathbf{N}}_{\mathcal{E}}^T \left(\tilde{\mathbf{N}}_{\mathcal{E}} \tilde{\mathbf{W}} \tilde{\mathbf{N}}_{\mathcal{E}}^T\right)^{-1} \tilde{\mathbf{N}}_{\mathcal{E}} \tilde{\mathbf{W}}\right) \tilde{\mathbf{N}}_{\mathcal{V}}^T \\
&= \tilde{\mathbf{N}}_{\mathcal{V}} \,\text{blkDiag}(\{\mathbf{T}_e\}_{e\in\mathcal{E}})\, \tilde{\mathbf{N}}_{\mathcal{V}}^T \\
&= \sum_{e\in\mathcal{E}} \tilde{\mathbf{N}}_{\mathcal{V}_e} \mathbf{T}_e \tilde{\mathbf{N}}_{\mathcal{V}_e}^T.
\end{aligned}$$

Each of these summands can be simplified to

$$\begin{aligned}
\tilde{\mathbf{N}}_{\mathcal{V}_e} \mathbf{T}_e \tilde{\mathbf{N}}_{\mathcal{V}_e}^T &= \tilde{\mathbf{N}}_{\mathcal{V}_e} \left(\frac{1}{h_e} \frac{1}{N_e} \left(N_e \mathbf{I}_{N_e} - \mathbf{1}_{N_e} \mathbf{1}_{N_e}^T\right)\right) \tilde{\mathbf{N}}_{\mathcal{V}_e}^T \\
&= \frac{1}{h_e} \tilde{\mathbf{N}}_{\mathcal{V}_e} \mathbf{I}_{N_e} \tilde{\mathbf{N}}_{\mathcal{V}_e}^T - \frac{1}{h_e N_e} \tilde{\mathbf{N}}_{\mathcal{V}_e} \left(\mathbf{1}_{N_e} \mathbf{1}_{N_e}^T\right) \tilde{\mathbf{N}}_{\mathcal{V}_e}^T \\
&= \frac{1}{h_e} \tilde{\mathbf{N}}_{\mathcal{V}_e} \tilde{\mathbf{N}}_{\mathcal{V}_e}^T - \frac{1}{\ell_e} \tilde{\mathbf{N}}_{\mathcal{V}_e} \mathbf{1}_{N_e} \left(\tilde{\mathbf{N}}_{\mathcal{V}_e} \mathbf{1}_{N_e}\right)^T
\end{aligned}$$

with

$$\begin{aligned}
\tilde{\mathbf{N}}_{\mathcal{V}_e} \tilde{\mathbf{N}}_{\mathcal{V}_e}^T &= \left(\mathbf{N}_e^{\text{out}} \otimes (\mathbf{e}_1^{N_e})^T + \mathbf{N}_e^{\text{in}} \otimes (\mathbf{e}_{N_e}^{N_e})^T\right) \left(\mathbf{N}_e^{\text{out}} \otimes (\mathbf{e}_1^{N_e})^T + \mathbf{N}_e^{\text{in}} \otimes (\mathbf{e}_{N_e}^{N_e})^T\right)^T \\
&= \mathbf{N}_e^{\text{out}} \mathbf{N}_e^{\text{out}^T} \otimes (\mathbf{e}_1^{N_e})^T \mathbf{e}_1^{N_e} + \mathbf{N}_e^{\text{out}} \mathbf{N}_e^{\text{in}^T} \otimes (\mathbf{e}_1^{N_e})^T \mathbf{e}_{N_e}^{N_e} \\
&\quad + \mathbf{N}_e^{\text{in}} \mathbf{N}_e^{\text{out}^T} \otimes (\mathbf{e}_{N_e}^{N_e})^T \mathbf{e}_1^{N_e} + \mathbf{N}_e^{\text{in}} \mathbf{N}_e^{\text{in}^T} \otimes (\mathbf{e}_{N_e}^{N_e})^T \mathbf{e}_{N_e}^{N_e} \\
&= \mathbf{N}_e^{\text{out}} \mathbf{N}_e^{\text{out}^T} + \mathbf{N}_e^{\text{in}} \mathbf{N}_e^{\text{in}^T}.
\end{aligned}$$

Here, $\mathbf{N}_e^{\text{out}}$ and $\mathbf{N}_e^{\text{in}}$ are vectors in $\mathbb{R}^n$ with only one non-zero entry -1 or 1 at the position of the origin $o(e)$ or terminal vertex $t(e)$ respectively. Thus, $\mathbf{N}_e^{\text{out}} \mathbf{N}_e^{\text{out}^T} + \mathbf{N}_e^{\text{in}} \mathbf{N}_e^{\text{in}^T}$ is a $n \times n$ diagonal matrix with two diagonal entries equal to 1 at $(t(e), t(e))$ and $(o(e), o(e))$. With this considerations, we observe that

$$\sum_{e\in\mathcal{E}} \frac{1}{h_e} \tilde{\mathbf{N}}_{\mathcal{V}_e} \tilde{\mathbf{N}}_{\mathcal{V}_e}^T$$

is exactly the weighted degree matrices of the original vertices in the extended graph, i.e.,

$$\sum_{e\in\mathcal{E}} \frac{1}{h_e} \tilde{\mathbf{N}}_{\mathcal{V}_e} \tilde{\mathbf{N}}_{\mathcal{V}_e}^T = \hat{\mathbf{L}}_{\mathcal{V}\mathcal{V}}.$$

This means that the Schur complement

$$\begin{aligned}
\hat{\mathbf{S}} &= \hat{\mathbf{L}}_{\mathcal{V}\mathcal{V}} - \hat{\mathbf{L}}_{\mathcal{V}\mathcal{E}} \hat{\mathbf{L}}_{\mathcal{E}\mathcal{E}}^{-1} \hat{\mathbf{L}}_{\mathcal{E}\mathcal{V}} \\
&= \hat{\mathbf{L}}_{\mathcal{V}\mathcal{V}} - \sum_{e\in\mathcal{E}} \tilde{\mathbf{N}}_{\mathcal{V}_e} \mathbf{T}_e \tilde{\mathbf{N}}_{\mathcal{V}_e}^T \\
&= \hat{\mathbf{L}}_{\mathcal{V}\mathcal{V}} - \left(\sum_{e\in\mathcal{E}} \frac{1}{h_e} \tilde{\mathbf{N}}_{\mathcal{V}_e} \tilde{\mathbf{N}}_{\mathcal{V}_e}^T - \sum_{e\in\mathcal{E}} \frac{1}{\ell_e} \tilde{\mathbf{N}}_{\mathcal{V}_e} \mathbf{1}_{N_e} \left(\tilde{\mathbf{N}}_{\mathcal{V}_e} \mathbf{1}_{N_e} \right)^T \right)
\end{aligned}$$

simplifies to the expression

$$\sum_{e\in\mathcal{E}} \frac{1}{\ell_e} \tilde{\mathbf{N}}_{\mathcal{V}_e} \mathbf{1}_{N_e} \left(\tilde{\mathbf{N}}_{\mathcal{V}_e} \mathbf{1}_{N_e} \right)^T.$$

Moreover, since

$$\tilde{\mathbf{N}}_{\mathcal{V}_e} \mathbf{1}_{N_e} = \mathbf{N}_e,$$

we obtain

$$\hat{\mathbf{S}} = \sum_{e\in\mathcal{E}} \frac{1}{\ell_e} \mathbf{N}_e \mathbf{N}_e^T,$$

which is the graph Laplacian matrix of the original graph with edge weights $w_e = \frac{1}{\ell_e}$. □

Lemma 3.11 *Under the assumptions of Theorem 3.10, it holds that*

$$\frac{1}{h_e} \tilde{\mathbf{N}}_e^T \left(\tilde{\mathbf{N}}_e \tilde{\mathbf{N}}_e^T \right)^{-1} \tilde{\mathbf{N}}_e = \frac{1}{h_e} \frac{1}{N_e} \left(N_e \mathbf{I}_{N_e} - \mathbf{1}_{N_e} \mathbf{1}_{N_e}^T \right),$$

where $\mathbf{1}_{N_e} \in \mathbb{R}^{N_e}$ denotes the vector of all ones.

Proof For the equilateral case, the assertion was proved in the context of a seminar paper.[3] The proof works equivalently for the general case. For the sake of completeness, the main steps are outlined here:

Since

$$\tilde{\mathbf{N}}_e = \begin{pmatrix} 1 & -1 & & & \\ & 1 & -1 & & \\ & & \ddots & \ddots & \\ & & & 1 & -1 \end{pmatrix} \in \mathbb{R}^{(N_e-1)\times N_e},$$

the inverse of the tridiagonal Toeplitz matrix

$$\tilde{\mathbf{N}}_e\tilde{\mathbf{N}}_e^T = \begin{pmatrix} 2 & -1 & & & \\ -1 & 2 & -1 & & \\ & \ddots & \ddots & \ddots & \\ & & -1 & 2 & -1 \\ & & & -1 & 2 \end{pmatrix} \in \mathbb{R}^{(N_e-1)\times(N_e-1)}$$

can be computed in closed form as

$$\left(\tilde{\mathbf{N}}_e\tilde{\mathbf{N}}_e^T\right)^{-1} = \frac{1}{N_e}\begin{pmatrix} N_e-1 & N_e-2 & \dots & 2 & 1 \\ N_e-2 & \ddots & \ddots & \ddots & 2 \\ \vdots & \ddots & \ddots & \ddots & \vdots \\ 2 & \ddots & \ddots & \ddots & N_e-2 \\ 1 & 2 & \dots & N_e-2 & N_e-1 \end{pmatrix},$$

compare [33]. Some matrix multiplication reveal

$$\begin{aligned}\frac{1}{h_e}\tilde{\mathbf{N}}_e^T\left(\tilde{\mathbf{N}}_e\tilde{\mathbf{N}}_e^T\right)^{-1}\tilde{\mathbf{N}}_e &= \frac{1}{h_e}\frac{1}{N_e}\begin{pmatrix} N_e-1 & -1 & \dots & -1 & -1 \\ -1 & \ddots & \ddots & \ddots & -1 \\ \vdots & \ddots & \ddots & \ddots & \vdots \\ -1 & \ddots & \ddots & \ddots & -1 \\ -1 & -1 & \dots & -1 & N_e-1 \end{pmatrix} \\ &= \frac{1}{h_e}\frac{1}{N_e}\left(N_e\mathbf{I} - \mathbf{1}_{N_e}\mathbf{1}_{N_e}^T\right).\end{aligned}$$

□

[3] Jaworski, R., Rhode, M., Schmidt, K.: Finite-Elemente-Methode fuer Quantum Graphen, Seminar paper in the summer semester 2022, Mathematical Institute, University of Cologne (2022), Lemma 7.5.

IMEX Scheme with Domain Decomposition Solver

An alternative to the multigrid method in Sect. 3.3.2 is to follow the example of [2] where systems involving the graph Laplacian of the extended graph are solved using the observed identity of the Schur complement of $\tilde{\mathbf{L}}$ and the discrete graph Laplacian matrix $\mathbf{L}$ ([2], Theorem 4.3). However, some more work has to be done here since we are dealing with non-equilateral graphs and the structure of the systems evolving from implicit time stepping methods differs from the systems arising from elliptic PDEs that are considered in [2]. We therefore derived the general Schur complement identity for non-equilateral graphs as

$$\hat{\mathbf{S}} := \hat{\mathbf{L}}/\hat{\mathbf{L}}_{\mathcal{E}\mathcal{E}} = \hat{\mathbf{L}}_{\Gamma}$$

in Theorem 3.10. It remains to deduce how this relation can also help us to solve systems of the form

$$(\hat{\mathbf{M}} + \Delta t\, \hat{\mathbf{L}})\mathbf{u}^{t+1} = \hat{\mathbf{M}}\mathbf{u}^t + \Delta t\, \hat{\mathbf{r}}(\mathbf{u}^t) =: \mathbf{b} \tag{3.27}$$

(compare (3.14)) efficiently.

The presented approach relies on the mass matrix approximation by the trapezoidal rule (compare Lemma 3.7), i.e., $\bar{\mathbf{M}} \approx \hat{\mathbf{M}}$ is diagonal. We may then write (3.27) as

$$(\mathbf{I} + \Delta t\, \bar{\mathbf{M}}^{-1}\hat{\mathbf{L}})\mathbf{u}^{t+1} = \mathbf{u}^t + \Delta t\, \bar{\mathbf{M}}^{-1}\hat{\mathbf{r}}(\mathbf{u}^t). \tag{3.28}$$

As the system on the left-hand side not symmetric, first observe that its solution is given by $\mathbf{u}^{t+1} = \bar{\mathbf{M}}^{-\frac{1}{2}}\mathbf{v}^{t+1}$ where $\mathbf{v}^{t+1}$ solves the symmetric system

$$\left(\mathbf{I} + \Delta t\, \bar{\mathbf{M}}^{-\frac{1}{2}}\hat{\mathbf{L}}\bar{\mathbf{M}}^{-\frac{1}{2}}\right)\mathbf{v}^{t+1} = \bar{\mathbf{M}}^{\frac{1}{2}}\mathbf{u}^t + \Delta t\, \bar{\mathbf{M}}^{-\frac{1}{2}}\hat{\mathbf{r}}(\mathbf{u}^t). \tag{3.29}$$

This can be seen by multiplying by $\bar{\mathbf{M}}^{\frac{1}{2}}$ from the left and expanding by $\mathbf{I} = \bar{\mathbf{M}}^{-\frac{1}{2}}\bar{\mathbf{M}}^{\frac{1}{2}}$. For ease of notation, we define $\mathbf{v} := \mathbf{v}^{t+1}$ and $\hat{\mathbf{b}} := \bar{\mathbf{M}}^{\frac{1}{2}}\mathbf{u}^t + \Delta t\, \bar{\mathbf{M}}^{-\frac{1}{2}}\hat{\mathbf{r}}(\mathbf{u}^t)$.

Since $\bar{\mathbf{M}}^{-\frac{1}{2}}$ is a diagonal matrix, it can be written as $\bar{\mathbf{M}}^{-\frac{1}{2}} = \mathrm{blkDiag}(\bar{\mathbf{M}}_{\mathcal{V}\mathcal{V}}^{-\frac{1}{2}}, \bar{\mathbf{M}}_{\mathcal{E}\mathcal{E}}^{-\frac{1}{2}})$ with two diagonal matrices $\bar{\mathbf{M}}_{\mathcal{V}\mathcal{V}}^{-\frac{1}{2}}$ and $\bar{\mathbf{M}}_{\mathcal{E}\mathcal{E}}^{-\frac{1}{2}}$ such that the symmetrized system assumes the block form

$$\begin{aligned}(\mathbf{I} + \Delta t\bar{\mathbf{M}}^{-\frac{1}{2}}\hat{\mathbf{L}}\bar{\mathbf{M}}^{-\frac{1}{2}})\mathbf{v} &= \begin{bmatrix} \mathbf{I} + \Delta t\, \bar{\mathbf{M}}_{\mathcal{V}\mathcal{V}}^{-\frac{1}{2}}\hat{\mathbf{L}}_{\mathcal{V}\mathcal{V}}\bar{\mathbf{M}}_{\mathcal{V}\mathcal{V}}^{-\frac{1}{2}} & \Delta t\, \bar{\mathbf{M}}_{\mathcal{V}\mathcal{V}}^{-\frac{1}{2}}\hat{\mathbf{L}}_{\mathcal{V}\mathcal{E}}\bar{\mathbf{M}}_{\mathcal{E}\mathcal{E}}^{-\frac{1}{2}} \\ \Delta t\, (\bar{\mathbf{M}}_{\mathcal{V}\mathcal{V}}^{-\frac{1}{2}}\hat{\mathbf{L}}_{\mathcal{V}\mathcal{E}}\bar{\mathbf{M}}_{\mathcal{E}\mathcal{E}}^{-\frac{1}{2}})^T & \mathbf{I} + \Delta t\, \bar{\mathbf{M}}_{\mathcal{E}\mathcal{E}}^{-\frac{1}{2}}\hat{\mathbf{L}}_{\mathcal{E}\mathcal{E}}\bar{\mathbf{M}}_{\mathcal{E}\mathcal{E}}^{-\frac{1}{2}} \end{bmatrix} \begin{bmatrix} \mathbf{v}_{\mathcal{V}} \\ \mathbf{v}_{\mathcal{E}} \end{bmatrix} \\ &= \begin{bmatrix} \hat{\mathbf{b}}_{\mathcal{V}} \\ \hat{\mathbf{b}}_{\mathcal{E}} \end{bmatrix}.\end{aligned}$$

The objective in the following is to consider a block factorization of this system, so we first prove the following lemma on the Schur complement.

Lemma 3.12 *Suppose we are given the diagonalization* $\bar{\mathbf{M}}_{\mathcal{E}\mathcal{E}}^{-\frac{1}{2}}\hat{\mathbf{L}}_{\mathcal{E}\mathcal{E}}\bar{\mathbf{M}}_{\mathcal{E}\mathcal{E}}^{-\frac{1}{2}} = \mathbf{U}\boldsymbol{\Lambda}\mathbf{U}^T$. *Then, the Schur complement of* $(\mathbf{I} + \Delta t\bar{\mathbf{M}}^{-\frac{1}{2}}\hat{\mathbf{L}}\bar{\mathbf{M}}^{-\frac{1}{2}})$ *can be expressed in terms of the original weighted graph Laplacian matrix as*

$$\hat{\mathbf{S}}_t = \Delta t\, \bar{\mathbf{M}}_{\mathcal{V}\mathcal{V}}^{-\frac{1}{2}}\hat{\mathbf{L}}_{\Gamma}\bar{\mathbf{M}}_{\mathcal{V}\mathcal{V}}^{-\frac{1}{2}} + \mathbf{I} - \Delta t^2 \left(\bar{\mathbf{M}}_{\mathcal{V}\mathcal{V}}^{-\frac{1}{2}}\hat{\mathbf{L}}_{\mathcal{V}\mathcal{E}}\bar{\mathbf{M}}_{\mathcal{E}\mathcal{E}}^{-\frac{1}{2}}\right)\left(\mathbf{U}\mathbf{X}^{-1}\mathbf{U}^T\right) \times \left(\bar{\mathbf{M}}_{\mathcal{V}\mathcal{V}}^{-\frac{1}{2}}\hat{\mathbf{L}}_{\mathcal{V}\mathcal{E}}\bar{\mathbf{M}}_{\mathcal{E}\mathcal{E}}^{-\frac{1}{2}}\right)^T$$

where $\mathbf{X}^{-1} := (\Delta t\boldsymbol{\Lambda} + \mathbf{I})^{-1} - (\Delta t\boldsymbol{\Lambda})^{-1}$.

Proof The Schur complement of the block matrix $(\mathbf{I} + \Delta t\bar{\mathbf{M}}^{-\frac{1}{2}}\hat{\mathbf{L}}\bar{\mathbf{M}}^{-\frac{1}{2}})$ is given by

$$\hat{\mathbf{S}}_t := \Delta t\left(\mathbf{I} + \Delta t\bar{\mathbf{M}}_{\mathcal{V}\mathcal{V}}^{-\frac{1}{2}}\hat{\mathbf{L}}_{\mathcal{V}\mathcal{V}}\bar{\mathbf{M}}_{\mathcal{V}\mathcal{V}}^{-\frac{1}{2}}\right) - \Delta t^2\left(\bar{\mathbf{M}}_{\mathcal{V}\mathcal{V}}^{-\frac{1}{2}}\hat{\mathbf{L}}_{\mathcal{V}\mathcal{E}}\bar{\mathbf{M}}_{\mathcal{E}\mathcal{E}}^{-\frac{1}{2}}\right)\left(\mathbf{I} + \Delta t\bar{\mathbf{M}}_{\mathcal{E}\mathcal{E}}^{-\frac{1}{2}}\hat{\mathbf{L}}_{\mathcal{E}\mathcal{E}}\bar{\mathbf{M}}_{\mathcal{E}\mathcal{E}}^{-\frac{1}{2}}\right)^{-1}\left(\bar{\mathbf{M}}_{\mathcal{V}\mathcal{V}}^{-\frac{1}{2}}\hat{\mathbf{L}}_{\mathcal{V}\mathcal{E}}\bar{\mathbf{M}}_{\mathcal{E}\mathcal{E}}^{-\frac{1}{2}}\right)^T.$$

In order to use the identity derived for the extended graph Laplacian Schur complement, we need to reformulate the inverse in the second summand of $\hat{\mathbf{S}}_t$. Consider therefore the diagonalization

$$\bar{\mathbf{M}}_{\mathcal{E}\mathcal{E}}^{-\frac{1}{2}}\hat{\mathbf{L}}_{\mathcal{E}\mathcal{E}}\bar{\mathbf{M}}_{\mathcal{E}\mathcal{E}}^{-\frac{1}{2}} = \mathbf{U}\boldsymbol{\Lambda}\mathbf{U}^T$$

which allows to reformulate

$$\begin{aligned}\left(\mathbf{I} + \Delta t\bar{\mathbf{M}}_{\mathcal{E}\mathcal{E}}^{-\frac{1}{2}}\hat{\mathbf{L}}_{\mathcal{E}\mathcal{E}}\bar{\mathbf{M}}_{\mathcal{E}\mathcal{E}}^{-\frac{1}{2}}\right)^{-1} &= \left(\mathbf{I} + \Delta t\mathbf{U}\boldsymbol{\Lambda}\mathbf{U}^T\right)^{-1} \\ &= \mathbf{U}\,(\mathbf{I} + \Delta t\boldsymbol{\Lambda})^{-1}\,\mathbf{U}^T\end{aligned}$$

since $\mathbf{U}\mathbf{U}^T = \mathbf{I}$. Defining $\mathbf{X}^{-1} := (\Delta t\boldsymbol{\Lambda} + \mathbf{I})^{-1} - (\Delta t\boldsymbol{\Lambda})^{-1}$, it holds that $(\mathbf{I} + \Delta t\boldsymbol{\Lambda})^{-1} = (\Delta t\boldsymbol{\Lambda})^{-1} + \mathbf{X}^{-1}$, and it follows

$$\begin{aligned}\left(\mathbf{I} + \Delta t\bar{\mathbf{M}}_{\mathcal{E}\mathcal{E}}^{-\frac{1}{2}}\hat{\mathbf{L}}_{\mathcal{E}\mathcal{E}}\bar{\mathbf{M}}_{\mathcal{E}\mathcal{E}}^{-\frac{1}{2}}\right)^{-1} &= \mathbf{U}\,(\mathbf{I} + \Delta t\boldsymbol{\Lambda})^{-1}\,\mathbf{U}^T \\ &= \mathbf{U}\left((\Delta t\boldsymbol{\Lambda})^{-1} + \mathbf{X}^{-1}\right)\mathbf{U}^T\end{aligned}$$

and consequently

$$\begin{aligned}
&\Delta t^2 \left(\bar{\mathbf{M}}_{\mathcal{VV}}^{-\frac{1}{2}}\hat{\mathbf{L}}_{\mathcal{VE}}\bar{\mathbf{M}}_{\mathcal{EE}}^{-\frac{1}{2}}\right)\left(\mathbf{I}+\Delta t\bar{\mathbf{M}}_{\mathcal{EE}}^{-\frac{1}{2}}\hat{\mathbf{L}}_{\mathcal{EE}}\bar{\mathbf{M}}_{\mathcal{EE}}^{-\frac{1}{2}}\right)^{-1}\left(\bar{\mathbf{M}}_{\mathcal{VV}}^{-\frac{1}{2}}\hat{\mathbf{L}}_{\mathcal{VE}}\bar{\mathbf{M}}_{\mathcal{EE}}^{-\frac{1}{2}}\right)^T \\
&\quad = \Delta t^2 \left(\bar{\mathbf{M}}_{\mathcal{VV}}^{-\frac{1}{2}}\hat{\mathbf{L}}_{\mathcal{VE}}\bar{\mathbf{M}}_{\mathcal{EE}}^{-\frac{1}{2}}\right)\left(\mathbf{U}\left((\Delta t\mathbf{\Lambda})^{-1}+\mathbf{X}^{-1}\right)\mathbf{U}^T\right)\left(\bar{\mathbf{M}}_{\mathcal{VV}}^{-\frac{1}{2}}\hat{\mathbf{L}}_{\mathcal{VE}}\bar{\mathbf{M}}_{\mathcal{EE}}^{-\frac{1}{2}}\right)^T \\
&\quad = \Delta t^2 \left(\bar{\mathbf{M}}_{\mathcal{VV}}^{-\frac{1}{2}}\hat{\mathbf{L}}_{\mathcal{VE}}\bar{\mathbf{M}}_{\mathcal{EE}}^{-\frac{1}{2}}\right)\left(\frac{1}{\Delta t}\mathbf{U}\mathbf{\Lambda}^{-1}\mathbf{U}^T+\mathbf{U}\mathbf{X}^{-1}\mathbf{U}^T\right)\left(\bar{\mathbf{M}}_{\mathcal{VV}}^{-\frac{1}{2}}\hat{\mathbf{L}}_{\mathcal{VE}}\bar{\mathbf{M}}_{\mathcal{EE}}^{-\frac{1}{2}}\right)^T.
\end{aligned}$$

Observe that $\mathbf{U}\mathbf{\Lambda}^{-1}\mathbf{U}^T = (\mathbf{U}\mathbf{\Lambda}\mathbf{U}^T)^{-1} = \left(\bar{\mathbf{M}}_{\mathcal{EE}}^{-\frac{1}{2}}\hat{\mathbf{L}}_{\mathcal{EE}}\bar{\mathbf{M}}_{\mathcal{EE}}^{-\frac{1}{2}}\right)^{-1}$ and hence

$$\begin{aligned}
\hat{\mathbf{S}}_t &= \left(\mathbf{I}+\Delta t\bar{\mathbf{M}}_{\mathcal{VV}}^{-\frac{1}{2}}\hat{\mathbf{L}}_{\mathcal{VV}}\bar{\mathbf{M}}_{\mathcal{VV}}^{-\frac{1}{2}}\right) \\
&\quad - \Delta t^2 \left(\bar{\mathbf{M}}_{\mathcal{VV}}^{-\frac{1}{2}}\hat{\mathbf{L}}_{\mathcal{VE}}\bar{\mathbf{M}}_{\mathcal{EE}}^{-\frac{1}{2}}\right)\left(\mathbf{I}+\Delta t\bar{\mathbf{M}}_{\mathcal{EE}}^{-\frac{1}{2}}\hat{\mathbf{L}}_{\mathcal{EE}}\bar{\mathbf{M}}_{\mathcal{EE}}^{-\frac{1}{2}}\right)^{-1}\left(\bar{\mathbf{M}}_{\mathcal{VV}}^{-\frac{1}{2}}\hat{\mathbf{L}}_{\mathcal{VE}}\bar{\mathbf{M}}_{\mathcal{EE}}^{-\frac{1}{2}}\right)^T \\
&= \Delta t\left(\bar{\mathbf{M}}_{\mathcal{VV}}^{-\frac{1}{2}}\hat{\mathbf{L}}_{\mathcal{VV}}\bar{\mathbf{M}}_{\mathcal{VV}}^{-\frac{1}{2}} - \left(\bar{\mathbf{M}}_{\mathcal{VV}}^{-\frac{1}{2}}\hat{\mathbf{L}}_{\mathcal{VE}}\bar{\mathbf{M}}_{\mathcal{EE}}^{-\frac{1}{2}}\right)\left(\bar{\mathbf{M}}_{\mathcal{EE}}^{-\frac{1}{2}}\hat{\mathbf{L}}_{\mathcal{EE}}\bar{\mathbf{M}}_{\mathcal{EE}}^{-\frac{1}{2}}\right)^{-1}\right. \\
&\quad \left.\times\left(\bar{\mathbf{M}}_{\mathcal{VV}}^{-\frac{1}{2}}\hat{\mathbf{L}}_{\mathcal{VE}}\bar{\mathbf{M}}_{\mathcal{EE}}^{-\frac{1}{2}}\right)^T\right) \\
&\quad + \mathbf{I} - \Delta t^2\left(\bar{\mathbf{M}}_{\mathcal{VV}}^{-\frac{1}{2}}\hat{\mathbf{L}}_{\mathcal{VE}}\bar{\mathbf{M}}_{\mathcal{EE}}^{-\frac{1}{2}}\right)\left(\mathbf{U}\mathbf{X}^{-1}\mathbf{U}^T\right)\left(\bar{\mathbf{M}}_{\mathcal{VV}}^{-\frac{1}{2}}\hat{\mathbf{L}}_{\mathcal{VE}}\bar{\mathbf{M}}_{\mathcal{EE}}^{-\frac{1}{2}}\right)^T.
\end{aligned}$$

We will now show that the first part of this expression can be reduced to the original graph similar to the Schur complement of $\hat{\mathbf{L}}$ discussed in Theorem 3.10. This is because

$$\begin{aligned}
&\bar{\mathbf{M}}_{\mathcal{VV}}^{-\frac{1}{2}}\hat{\mathbf{L}}_{\mathcal{VV}}\bar{\mathbf{M}}_{\mathcal{VV}}^{-\frac{1}{2}} - \left(\bar{\mathbf{M}}_{\mathcal{VV}}^{-\frac{1}{2}}\hat{\mathbf{L}}_{\mathcal{VE}}\bar{\mathbf{M}}_{\mathcal{EE}}^{-\frac{1}{2}}\right)\left(\bar{\mathbf{M}}_{\mathcal{EE}}^{-\frac{1}{2}}\hat{\mathbf{L}}_{\mathcal{EE}}\bar{\mathbf{M}}_{\mathcal{EE}}^{-\frac{1}{2}}\right)^{-1}\left(\bar{\mathbf{M}}_{\mathcal{VV}}^{-\frac{1}{2}}\hat{\mathbf{L}}_{\mathcal{VE}}\bar{\mathbf{M}}_{\mathcal{EE}}^{-\frac{1}{2}}\right)^T \\
&\quad = \bar{\mathbf{M}}_{\mathcal{VV}}^{-\frac{1}{2}}\hat{\mathbf{L}}_{\mathcal{VV}}\bar{\mathbf{M}}_{\mathcal{VV}}^{-\frac{1}{2}} - \left(\bar{\mathbf{M}}_{\mathcal{VV}}^{-\frac{1}{2}}\hat{\mathbf{L}}_{\mathcal{VE}}\bar{\mathbf{M}}_{\mathcal{EE}}^{-\frac{1}{2}}\right)\left(\bar{\mathbf{M}}_{\mathcal{EE}}^{\frac{1}{2}}\hat{\mathbf{L}}_{\mathcal{EE}}^{-1}\bar{\mathbf{M}}_{\mathcal{EE}}^{\frac{1}{2}}\right)\left(\bar{\mathbf{M}}_{\mathcal{EE}}^{-\frac{1}{2}}\hat{\mathbf{L}}_{\mathcal{VE}}^T\bar{\mathbf{M}}_{\mathcal{VV}}^{-\frac{1}{2}}\right) \\
&\quad = \bar{\mathbf{M}}_{\mathcal{VV}}^{-\frac{1}{2}}\hat{\mathbf{L}}_{\mathcal{VV}}\bar{\mathbf{M}}_{\mathcal{VV}}^{-\frac{1}{2}} - \bar{\mathbf{M}}_{\mathcal{VV}}^{-\frac{1}{2}}\hat{\mathbf{L}}_{\mathcal{VE}}\hat{\mathbf{L}}_{\mathcal{EE}}^{-1}\hat{\mathbf{L}}_{\mathcal{VE}}^T\bar{\mathbf{M}}_{\mathcal{VV}}^{-\frac{1}{2}} \\
&\quad = \bar{\mathbf{M}}_{\mathcal{VV}}^{-\frac{1}{2}}(\hat{\mathbf{L}}_{\mathcal{VV}} - \hat{\mathbf{L}}_{\mathcal{VE}}\hat{\mathbf{L}}_{\mathcal{EE}}^{-1}\hat{\mathbf{L}}_{\mathcal{VE}}^T)\bar{\mathbf{M}}_{\mathcal{VV}}^{-\frac{1}{2}}.
\end{aligned}$$

The expression in the middle of the last row is exactly the Schur complement of $\hat{\mathbf{L}}$. According to Theorem 3.10, it is equal to $\hat{\mathbf{L}}_{\Gamma}$. □

With this observations in place, we now consider the factorization of $(\mathbf{I} + \Delta t \bar{\mathbf{M}}^{-\frac{1}{2}} \hat{\mathbf{L}} \bar{\mathbf{M}}^{-\frac{1}{2}})$ given by

$$\begin{bmatrix} \hat{\mathbf{S}}_t & \Delta t \left(\bar{\mathbf{M}}_{\mathcal{V}\mathcal{V}}^{-\frac{1}{2}} \hat{\mathbf{L}}_{\mathcal{V}\mathcal{E}} \bar{\mathbf{M}}_{\mathcal{E}\mathcal{E}}^{-\frac{1}{2}} \right) \\ \mathbf{0} & \mathbf{I} + \Delta t \left(\bar{\mathbf{M}}_{\mathcal{E}\mathcal{E}}^{-\frac{1}{2}} \hat{\mathbf{L}}_{\mathcal{E}\mathcal{E}} \bar{\mathbf{M}}_{\mathcal{E}\mathcal{E}}^{-\frac{1}{2}} \right) \end{bmatrix}$$
$$\times \begin{bmatrix} \mathbf{I} & \mathbf{0} \\ \left(\mathbf{I} + \Delta t \bar{\mathbf{M}}_{\mathcal{E}\mathcal{E}}^{-\frac{1}{2}} \hat{\mathbf{L}}_{\mathcal{E}\mathcal{E}} \bar{\mathbf{M}}_{\mathcal{E}\mathcal{E}}^{-\frac{1}{2}} \right)^{-1} \Delta t \left(\bar{\mathbf{M}}_{\mathcal{V}\mathcal{V}}^{-\frac{1}{2}} \hat{\mathbf{L}}_{\mathcal{V}\mathcal{E}} \bar{\mathbf{M}}_{\mathcal{E}\mathcal{E}}^{-\frac{1}{2}} \right)^T & \mathbf{I} \end{bmatrix}$$

and apply a block elimination procedure to the system $(\mathbf{I} + \Delta t \bar{\mathbf{M}}^{-\frac{1}{2}} \hat{\mathbf{L}} \bar{\mathbf{M}}^{-\frac{1}{2}})\mathbf{v} = \hat{\mathbf{b}}$. This leads to two reduced systems

$$\hat{\mathbf{S}}_t \mathbf{v}_{\mathcal{V}} = \hat{\mathbf{b}}_{\mathcal{V}} - \Delta t \left(\bar{\mathbf{M}}_{\mathcal{V}\mathcal{V}}^{-\frac{1}{2}} \hat{\mathbf{L}}_{\mathcal{V}\mathcal{E}} \bar{\mathbf{M}}_{\mathcal{E}\mathcal{E}}^{-\frac{1}{2}} \right) \times \left(\mathbf{I} + \Delta t \bar{\mathbf{M}}_{\mathcal{E}\mathcal{E}}^{-\frac{1}{2}} \hat{\mathbf{L}}_{\mathcal{E}\mathcal{E}} \bar{\mathbf{M}}_{\mathcal{E}\mathcal{E}}^{-\frac{1}{2}} \right)^{-1} \hat{\mathbf{b}}_{\mathcal{E}}$$

$$\left(\mathbf{I} + \Delta t \, \bar{\mathbf{M}}_{\mathcal{E}\mathcal{E}}^{-\frac{1}{2}} \hat{\mathbf{L}}_{\mathcal{E}\mathcal{E}} \bar{\mathbf{M}}_{\mathcal{E}\mathcal{E}}^{-\frac{1}{2}} \right) \mathbf{v}_{\mathcal{E}} = \hat{\mathbf{b}}_{\mathcal{E}} - \Delta t \left(\bar{\mathbf{M}}_{\mathcal{V}\mathcal{V}}^{-\frac{1}{2}} \hat{\mathbf{L}}_{\mathcal{V}\mathcal{E}} \bar{\mathbf{M}}_{\mathcal{E}\mathcal{E}}^{-\frac{1}{2}} \right)^T \mathbf{v}_{\mathcal{V}}.$$

The eigendecomposition $\bar{\mathbf{M}}_{\mathcal{E}\mathcal{E}}^{-\frac{1}{2}} \hat{\mathbf{L}}_{\mathcal{E}\mathcal{E}} \bar{\mathbf{M}}_{\mathcal{E}\mathcal{E}}^{-\frac{1}{2}} = \mathbf{U}\mathbf{\Lambda}\mathbf{U}^T$ is already given, i.e., the inverse appearing in the right hand side of the first system can be readily computed as

$$(\mathbf{I} + \Delta t \bar{\mathbf{M}}_{\mathcal{E}\mathcal{E}}^{-\frac{1}{2}} \hat{\mathbf{L}}_{\mathcal{E}\mathcal{E}} \bar{\mathbf{M}}_{\mathcal{E}\mathcal{E}}^{-\frac{1}{2}})^{-1} = \mathbf{U}(\mathbf{I} + \Delta t \mathbf{\Lambda})^{-1} \mathbf{U}^T. \tag{3.30}$$

By this, also the solution of the second system follows straightaway. The solution of the first system involving the Schur complement (which is of size $n \times n$!) can be efficiently computed for example by the CG algorithm.

The findings are summarized in the following solution scheme:

Remark 3.13 The computationally most expensive part of Algorithm 3 is the eigendecomposition in (i). Indeed, the matrix $\bar{\mathbf{M}}_{\mathcal{E}\mathcal{E}}^{-\frac{1}{2}} \hat{\mathbf{L}}_{\mathcal{E}\mathcal{E}} \bar{\mathbf{M}}_{\mathcal{E}\mathcal{E}}^{-\frac{1}{2}}$ is a block diagonal matrix of the form

$$\bar{\mathbf{M}}_{\mathcal{E}\mathcal{E}}^{-\frac{1}{2}} \hat{\mathbf{L}}_{\mathcal{E}\mathcal{E}} \bar{\mathbf{M}}_{\mathcal{E}\mathcal{E}}^{-\frac{1}{2}} = \operatorname{blkDiag} \left(\bar{\mathbf{M}}_{e_1}^{-\frac{1}{2}} \hat{\mathbf{L}}_{e_1} \bar{\mathbf{M}}_{e_1}^{-\frac{1}{2}}, \ldots, \bar{\mathbf{M}}_{e_m}^{-\frac{1}{2}} \hat{\mathbf{L}}_{e_m} \bar{\mathbf{M}}_{e_m}^{-\frac{1}{2}} \right)$$

where each block corresponds to an edge. Since the step lengths are constant across a fixed edge e, the blocks $\bar{\mathbf{M}}_{e_1}^{-\frac{1}{2}} \hat{\mathbf{L}}_{e_1} \bar{\mathbf{M}}_{e_1}^{-\frac{1}{2}}$ are Toeplitz matrices, i.e., $\mathbf{\Lambda}_e$ and $\mathbf{U}_e$ can be

Algorithm 3 Domain decomposition solver

(i) Compute the Eigendecomposition $\bar{\mathbf{M}}_{\mathcal{E}\mathcal{E}}^{-\frac{1}{2}}\hat{\mathbf{L}}_{\mathcal{E}\mathcal{E}}\bar{\mathbf{M}}_{\mathcal{E}\mathcal{E}}^{-\frac{1}{2}} = \mathbf{U}\boldsymbol{\Lambda}\mathbf{U}^T$

(ii) For each time step $t = 0, \ldots T$ set $\bar{\mathbf{M}}^{\frac{1}{2}}\mathbf{u}^t =: \hat{\mathbf{b}}$ and solve $\left(\mathbf{I} + \Delta t\,\bar{\mathbf{M}}^{-\frac{1}{2}}\hat{\mathbf{L}}\bar{\mathbf{M}}^{-\frac{1}{2}}\right)\mathbf{v} = \hat{\mathbf{b}}$ by solving the two reduced systems

(ii.1) $\hat{\mathbf{S}}_t\mathbf{v}_{\mathcal{V}} = \hat{\mathbf{b}}_{\mathcal{V}} - \Delta t\left(\bar{\mathbf{M}}_{\mathcal{V}\mathcal{V}}^{-\frac{1}{2}}\hat{\mathbf{L}}_{\mathcal{V}\mathcal{E}}\bar{\mathbf{M}}_{\mathcal{E}\mathcal{E}}^{-\frac{1}{2}}\right)\left(\mathbf{I} + \Delta t\bar{\mathbf{M}}_{\mathcal{E}\mathcal{E}}^{-\frac{1}{2}}\hat{\mathbf{L}}_{\mathcal{E}\mathcal{E}}\bar{\mathbf{M}}_{\mathcal{E}\mathcal{E}}^{-\frac{1}{2}}\right)^{-1}\hat{\mathbf{b}}_{\mathcal{E}}$ with

$$\hat{\mathbf{S}}_t := \Delta t\bar{\mathbf{M}}_{\mathcal{V}\mathcal{V}}^{-\frac{1}{2}}\hat{\mathbf{L}}_{\Gamma}\bar{\mathbf{M}}_{\mathcal{V}\mathcal{V}}^{-\frac{1}{2}} + \mathbf{I} - \Delta t^2\left(\bar{\mathbf{M}}_{\mathcal{V}\mathcal{V}}^{-\frac{1}{2}}\hat{\mathbf{L}}_{\mathcal{V}\mathcal{E}}\bar{\mathbf{M}}_{\mathcal{E}\mathcal{E}}^{-\frac{1}{2}}\right)\left(\mathbf{U}\mathbf{X}^{-1}\mathbf{U}^T\right)\left(\bar{\mathbf{M}}_{\mathcal{V}\mathcal{V}}^{-\frac{1}{2}}\hat{\mathbf{L}}_{\mathcal{V}\mathcal{E}}\bar{\mathbf{M}}_{\mathcal{E}\mathcal{E}}^{-\frac{1}{2}}\right)^T$$

and $\mathbf{X}^{-1} := (\Delta t\boldsymbol{\Lambda} + \mathbf{I})^{-1} - (\Delta t\boldsymbol{\Lambda})^{-1}$

(ii.2) $\left(\mathbf{I} + \Delta t\bar{\mathbf{M}}_{\mathcal{E}\mathcal{E}}^{-\frac{1}{2}}\hat{\mathbf{L}}_{\mathcal{E}\mathcal{E}}\bar{\mathbf{M}}_{\mathcal{E}\mathcal{E}}^{-\frac{1}{2}}\right)\mathbf{v}_{\mathcal{E}} = \hat{\mathbf{b}}_{\mathcal{E}} - \Delta t\left(\bar{\mathbf{M}}_{\mathcal{V}\mathcal{V}}^{-\frac{1}{2}}\hat{\mathbf{L}}_{\mathcal{V}\mathcal{E}}\bar{\mathbf{M}}_{\mathcal{E}\mathcal{E}}^{-\frac{1}{2}}\right)^T\mathbf{v}_{\mathcal{V}}$

and compute $\mathbf{u}^{t+1} = \bar{\mathbf{M}}^{-\frac{1}{2}}\mathbf{v}$.

computed in closed form for each block. However, these are dense matrices which means that in particular the assembling of (3.30) is expensive although it can be performed parallel for each edge. An alternative to the direct assembling is to solve

$$\text{(ii.1}') \qquad \hat{\mathbf{S}}_t\mathbf{v}_{\mathcal{V}} = \hat{\mathbf{b}}_{\mathcal{V}} - \Delta t\left(\bar{\mathbf{M}}_{\mathcal{V}\mathcal{V}}^{-\frac{1}{2}}\hat{\mathbf{L}}_{\mathcal{V}\mathcal{E}}\bar{\mathbf{M}}_{\mathcal{E}\mathcal{E}}^{-\frac{1}{2}}\right)\hat{\hat{\mathbf{b}}}_{\mathcal{E}}$$

where $\hat{\hat{\mathbf{b}}}_{\mathcal{E}}$ is obtained from a tridiagonal solver as the solution of

$$\left(\mathbf{I} + \Delta t\bar{\mathbf{M}}_{\mathcal{E}\mathcal{E}}^{-\frac{1}{2}}\hat{\mathbf{L}}_{\mathcal{E}\mathcal{E}}\bar{\mathbf{M}}_{\mathcal{E}\mathcal{E}}^{-\frac{1}{2}}\right)\hat{\hat{\mathbf{b}}}_{\mathcal{E}} = \hat{\mathbf{b}}_{\mathcal{E}}.$$

Again, the systems are completely decoupled such that the solution of the tridiagonal system can be conducted independently and in parallel for each edge.

If the system (ii.1) is solved by a direct solver, Algorithm 3 is itself a direct solver for (3.28) and no additional error is introduced. However, we have to keep in mind that we have chosen the diagonal approximation $\bar{\mathbf{M}}$ of the Mass matrix for the derivation of the domain decomposition scheme. Yet, the numerical findings in Sect. 6.1 suggest that the additional error introduced is negligible and, in particular, does not influence the order of convergence.

Chapter 4
Spectral Solution Method

Abstract In the present chapter, we derive a spectral Galerkin semidiscretization of Problems 2.21 and 2.22 in which we choose the eigenfunctions of the operator $\mathcal{H}$ as trial functions. This is preceded by an investigation of the eigenfunctions in view of their approximation properties. In particular, an eigenvalue estimate derived from Weyl's law allows an estimation of the truncation error. Further, we discuss the approximation error of the spectral Galerkin semidiscretization as well as the solution of the evolving system of ODEs by exponential integrators. Finally, we examine two particular aspects of implementation and give an outlook on the application to IBVP governed by other classes of PDEs, such as a fractional diffusion equation.

To our best knowledge, the only solution approach exploring quantum graph eigenfunctions so far was proposed by [10] which we reviewed in the introduction. However, the idea of developing a spectral method here emerged independent of their work and has been partly published in [1]. The main novelty of the following presentation is the structured derivation of the truncation error, the class of functions for which spectral convergence can be obtained, the derivation of error estimates, the application to semilinear equations as well as the efficient computation of inner products. Some of the results have been prepared in joint work with Prof. Dr. Mark Ainsworth for an internal manuscript.[1]

[1] Ainsworth, M., Weller, A.: A spectral Galerkin method for reaction-diffusion equations on metric graphs [unpublished manuscript], 2021.

A. Weller, *Numerical Methods for Metric Graphs*, Lecture Notes in Mathematics 2382, https://doi.org/10.1007/978-3-032-05011-3_4

4.1 Trial Functions

The spectral method relies on a Galerkin discretization, i.e., we solve the weak formulation on a finite dimensional subspace. As trial functions, we choose linear combinations of eigenfunctions of the differential operator $\mathcal{H}$ as introduced in the following.

4.1.1 Eigenfunction Expansion

The spectrum of a quantum graph is identified with the spectrum of $\mathcal{H}$ acting on functions on Γ equipped with Neumann-Kirchhoff boundary conditions. Throughout this book, we only consider the standard differential operator (2.1) and Neumann-Kirchhoff conditions (2.2), i.e., the spectrum of the quantum graph only depends on the metric graph Γ in the sense of the following definition.

Definition 4.1 Let Γ be a metric graph. We define the *spectrum of* Γ, denoted by $\sigma(\Gamma)$, as the spectrum of $\mathcal{H} : u \mapsto -\frac{\mathrm{d}^2 u}{\mathrm{d}x^2}$ acting on $\mathrm{dom}_{\mathcal{H},\mathrm{NK}}$ (see Definition 2.13).

We summarize the following general result on the spectrum of compact metric graphs, which can be found for example in [6] and [34]:

Theorem 4.2 *For a compact metric graph* Γ*, the spectrum* $\sigma(\Gamma)$ *is completely determined by eigenvalues* $\lambda \in \mathbb{R}$ *for which a nontrivial* ϕ *with*

$$\mathcal{H}\phi = \lambda\phi \tag{4.1}$$

such that ϕ *fulfills the Neumann-Kirchhoff conditions* (2.2) *exists. In particular,* $\lambda \geq 0$ *and the corresponding eigensolutions are referred to as eigenfunctions, denoted by* ϕ*.*

Proof In [6], Theorem 3.1.1., the authors prove that the spectrum $\sigma(\Gamma)$ is composed of an ascending sequence of eigenvalues λ_q of finite multiplicity with $\lim_{q\to\infty} \lambda_q = \infty$. Additionally, $\mathcal{H}$ is positive-semidefinite since (2.12) is non-negative. Together with the self-adjointness, we obtain that $\lambda \in \mathbb{R}$ and $\lambda \geq 0$. □

Since $\mathcal{H} : u \mapsto -\frac{\mathrm{d}^2 u}{\mathrm{d}x^2}$ with Neumann-Kirchhoff conditions is self-adjoint, we can expand $u \in L^2(\Gamma)$ in terms of the (orthogonal) eigenfunctions ϕ_λ, $\lambda \in \sigma(\Gamma)$ of $\mathcal{H}$ as

$$u = \sum_{\lambda} c_\lambda \phi_\lambda \tag{4.2}$$

with coefficients $c_\lambda \in \mathbb{R}$. The idea of the spectral solution approach is to truncate the (infinite) series in (4.2) such that we obtain a finite dimensional approximation.

We will from now on always assume the eigenfunctions to be normalized, i.e., we consider an orthonormal basis $\varphi := \{\phi_\lambda, \lambda \in \sigma(\Gamma)\}$. For convenience, we will enumerate the eigenvalues in ascending order and in a slight abuse of notation write λ_q for the q-th eigenvalue with associated eigenfunction ϕ_q. The basis φ can thus be equivalently written as $\varphi = \{\phi_q, q = 1, 2, \ldots\}$ and we denote the space spanned by this basis by X.

In the next chapter, we will derive in detail an efficient method to compute an arbitrary number of eigenfunctions numerically. At some points in the present chapter, however, we have to anticipate that the eigenfunctions can be expressed on each edge e as

$$\phi_q(x) = A_e^q \cos\left(\sqrt{\lambda_q}\, x\right) + B_e^q \sin\left(\sqrt{\lambda_q}\, x\right)$$

with some edge specific constants A_e^q, B_e^q.

4.1.2 Approximation Theory

As our objective is to truncate the series in (4.2), we are interested in a finite dimensional subspace of X spanned by a finite number of eigenfunctions $\varphi_Q := \{\phi_q, q = 1, \ldots, Q\}$ for $Q \in \mathbb{N}$, i.e.,

$$X_Q := \operatorname{span}\{\phi_q, q = 1, \ldots, Q\}.$$

In particular, we first define the following L^2-projection onto X_Q.

Theorem 4.3 *$P_Q : L^2(\Gamma) \to X_Q$ with*

$$P_Q u := \sum_{q \le Q} \left(u, \phi_q\right)_\Gamma \phi_q \tag{4.3}$$

defines an L^2-projection of $u \in L^2(\Gamma)$ onto $X_Q = \operatorname{span}\{\phi_q, q = 1, \ldots, Q\}$.

Proof Representing $u \in L^2(\Gamma)$ as $u = \sum_q c_q \phi_q$ with constants $c_q \in \mathbb{R}$ yields

$$(u, \phi)_\Gamma = \left(\sum_q c_q \phi_q, \phi\right)_\Gamma = \sum_q c_q \left(\phi_q, \phi\right)_\Gamma$$

for all $\phi \in X_Q$. All $\phi \in X$ are orthonormal, i.e., for a fixed $\phi_q \in X_Q$, all terms but $c_q\left(\phi_q, \phi_q\right)_\Gamma$ in the sum vanish and, in particular, $\left(\phi_q, \phi_q\right)_\Gamma = 1$. We conclude that the coefficients c_q are given by

$$c_q = \left(u, \phi_q\right)_\Gamma.$$

Recall that an L^2-projection $P_Q : L^2(\Gamma) \to X_Q$ onto X_Q is defined by

$$\left(u - P_Q u, \phi\right)_\Gamma = 0 \quad \text{for all } \phi \in X_Q$$

which is fulfilled for $P_Q u := \sum_{q \le Q} \left(u, \phi_q\right)_\Gamma \phi_q$ given in the assertion as the orthogonality implies

$$\begin{aligned}\left(u - P_Q u, \phi\right)_\Gamma &= \left(\sum_q \left(u, \phi_q\right)_\Gamma \phi_q - \sum_{q \le Q} \left(u, \phi_q\right)_\Gamma \phi_q, \phi\right)_\Gamma \\ &= \left(\sum_{q>Q} c_q \phi_q, \phi\right)_\Gamma = 0\end{aligned}$$

for all $\phi \in X_Q$. □

We are confronted with the question of how well a function $u = \sum_q c_q \phi_q$ can be approximated by the projection $P_Q u$. Following a standard approach (compare for example [12]), we use Parseval's identity to express the *truncation error* as

$$\left\|u - P_Q u\right\|_\Gamma^2 = \sum_{q>Q} \left|\left(u, \phi_q\right)_\Gamma\right|^2 = \sum_{q>Q} |c_q|^2. \tag{4.4}$$

In other words, the error is essentially determined by the *decay of coefficients* c_q as $q \to \infty$. In [10], the authors attempt to estimate the rate of decay and claim that *spectral accuracy* can only be obtained for sufficiently smooth functions with compact support on each edge. But considering functions with compact support on each edge is the same as posing zero Dirichlet conditions at the vertices of the graph instead of Neumann-Kirchhoff conditions. This in turn uncouples the system to m independent equations on the edges and the graph structure has no influence on the solution at all. We will therefore show that the applicable class of functions can be further extended. Let us first define $\mathcal{H}^{(d)} u := (-1)^d u^{(2d)} := (-1)^d \frac{\mathrm{d}^{2d}}{\mathrm{d}x^{2d}} u$ and

$$\begin{aligned}\mathrm{dom}^d_{\mathcal{H},\mathrm{NK}} := \bigoplus_{e \in \mathcal{E}} H^{2d}(e) \cap \{&u, u^{(2)}, \ldots, u^{(2d)} \text{ fulfill the Neumann-Kirchhoff} \\ &\text{conditions}\}.\end{aligned}$$

In other words, with $u \in \mathrm{dom}^d_{\mathcal{H},\mathrm{NK}}$, we mean that u is sufficiently smooth and all even derivatives are in $\mathrm{dom}_{\mathcal{H},\mathrm{NK}}$.

Theorem 4.4 *For* $u \in \mathrm{dom}^d_{\mathcal{H},\mathrm{NK}}$, $q > 1$, *the projection coefficients are bounded by*

$$|c_q| \le \frac{1}{\lambda_q^d} \left\|u^{(2d)}\right\|_\Gamma.$$

Proof Since $\lambda_1 = 0$ is a simple eigenvalue, it suffices to consider λ_q for $q > 1$ in the following. The proof essentially follows the arguments in [12], Section 5.2.1. The coefficients c_q of the expansion $\sum_q c_q \phi_q$ are given by $c_q = \left(u, \phi_q\right)_\Gamma$. Throughout this proof, we will apply the eigenvalue identity in the form $\phi_q = \frac{1}{\lambda_q}\mathcal{H}\phi_q$ and the identity

$$(\mathcal{H}u, \phi)_\Gamma = (u, \mathcal{H}\phi)_\Gamma$$

for $u, \phi \in \mathrm{dom}_{\mathcal{H},\mathrm{NK}}$ which follows from the self-adjointness of $\mathcal{H}$. It is then straightforward to derive

$$c_q = \left(u, \phi_q\right)_\Gamma = \frac{1}{\lambda_q}\left(u, \mathcal{H}\phi_q\right)_\Gamma = \frac{1}{\lambda_q}(\mathcal{H}u, \phi_q)_\Gamma.$$

If $\mathcal{H}u \in \mathrm{dom}_{\mathcal{H},\mathrm{NK}}$, we can repeat these arguments and arrive at

$$|c_q| = \left|\frac{1}{\lambda_q^2}\left(u^{(4)}, \phi_q\right)_\Gamma\right|$$

and for $u \in \mathrm{dom}^d_{\mathcal{H},\mathrm{NK}}$ finally at

$$|c_q| = \left|\frac{1}{\lambda_q^d}\left(u^{(2d)}, \phi_q\right)_\Gamma\right|. \tag{4.5}$$

The assertion follows by

$$|c_q| = \frac{1}{\lambda_q^d}\left|\left(u^{(2d)}, \phi_q\right)_\Gamma\right| \le \frac{1}{\lambda_q^d}\left\|u^{(2d)}\right\|_\Gamma \|\phi_q\|_\Gamma = \frac{1}{\lambda_q^d}\left\|u^{(2d)}\right\|_\Gamma$$

since we consider normalized eigenfunctions. □

The truncation error then follows as a direct consequence.

Corollary 4.5 *Under the conditions of Theorem 4.4, the truncation error fulfills*

$$\left\|u - P_Q u\right\|_\Gamma \le \frac{1}{\lambda_Q^d}\left\|u^{(2d)}\right\|_\Gamma.$$

Proof From (4.4) we deduce

$$\left\|u - P_Q u\right\|_\Gamma = \left(\sum_{q>Q} |c_q|^2\right)^{\frac{1}{2}} = \left(\sum_{q>Q} \frac{1}{\lambda_q^{2d}} \lambda_q^{2d} |c_q|^2\right)^{\frac{1}{2}} \le \frac{1}{\lambda_Q^d}\left(\sum_{q>Q} \lambda_q^{2d} |c_q|^2\right)^{\frac{1}{2}}$$

and thus by (4.5), the eigenvalue identity and the self-adjointness

$$\left\|u - P_Q u\right\|_\Gamma \leq \frac{1}{\lambda_Q^d}\left(\sum_{q>Q} \lambda_q^{2d} \left|c_q\right|^2\right)^{\frac{1}{2}} = \frac{1}{\lambda_Q^d}\left(\sum_{q>Q} \left|\left(u^{(2d)}, \phi_q\right)_\Gamma\right|^2\right)^{\frac{1}{2}}.$$

Since $\left(u^{(2d)}, \phi_q\right)_\Gamma$ are the projection coefficients of $u^{(2d)}$, Parseval's identity (applied to $u^{(2d)}$) yields

$$\left\|u^{(2d)}\right\|_\Gamma = \sum_q \left|\left(u^{2d}, \phi_q\right)_\Gamma\right|^2 \geq \sum_{q>Q} \left|\left(u^{(2d)}, \phi_q\right)_\Gamma\right|^2.$$

Together, we thus have $\left\|u - P_Q u\right\|_\Gamma \leq \frac{1}{\lambda_Q^d}\left\|u^{(2d)}\right\|_\Gamma$. □

4.1.3 Eigenvalue Estimates

The following result on eigenvalue counting, which we cite from [5], Lemma 4.4., can be used to derive a lower bound on the q-th eigenvalue.[2]

Theorem 4.6 *Given an eigenvalue λ_q, the number of eigenvalues smaller than or equal to λ_q^2 is given by*

$$\left|\{\lambda \in \sigma(\Gamma) : \lambda \leq \lambda_q^2\}\right| = \frac{\mathrm{vol}_\Gamma}{\pi}\lambda_q + O(1)$$

where vol_Γ *is the total length of the graph and* $O(1)$ *is independent of* λ_q.

Such results on eigenvalue counting are often referred to as *Weyl's Law*. In particular, the remainder term $O(1)$ can be bounded from below and above such that we obtain

$$\frac{\mathrm{vol}_\Gamma}{\pi}\lambda_q - m \leq \left|\{\lambda \in \sigma(\Gamma) : \lambda \leq \lambda_q^2\right| \leq \frac{\mathrm{vol}_\Gamma}{\pi}\lambda_q + n$$

where n is the number of vertices and m the number of edges of Γ. In other words, for q sufficiently large, the q-th eigenvalue λ_q can be bounded from below by

$$\lambda_q \geq \left(\frac{(q-n)\pi}{\mathrm{vol}_\Gamma}\right)^2.$$

[2] Note that this result has also been used in [10] to estimate the asymptotic behavior of the eigenvalues.

As observed in [10], we will see that, in practice, the eigenvalues often follow *Weyl's slope* $\left(\frac{q\pi}{\mathrm{vol}_\Gamma}\right)^2$ much more closely. We will address this further in the numerical experiments in Sect. 6.3.2.

Theorem 4.6 implies spectral accuracy since it allows to estimate the rate of decay of the coefficients in the spectral expansion as follows.

Corollary 4.7 *Under the conditions of Theorem 4.4, the projection coefficients are bounded by*

$$|c_Q| \leq \bar{C}\left(\frac{\mathrm{vol}_\Gamma}{(Q-n)\pi}\right)^{2d}\left\|u^{(2d)}\right\|_\Gamma.$$

with $\bar{C} > 0$.

4.2 Spectral Galerkin Approximation

We are now in the position to formulate the spectral Galerkin approximation of the generalized heat equation in weak form (Problem 2.21): Find $\boldsymbol{u}_Q \in L^2([0,T]; X_Q)$ with

$$\begin{aligned} \left(\boldsymbol{u}_Q'(t), g_Q\right)_\Gamma + \mathfrak{h}(\boldsymbol{u}_Q(t), g_Q) &= \left(\boldsymbol{f}(t), g_Q\right)_\Gamma \quad \text{for all} \quad g_Q \in X_Q,\ t > 0 \\ \boldsymbol{u}_Q(0) &= u_Q^0 \end{aligned} \tag{4.6}$$

where u_Q^0 is an approximation of the initial condition u^0 on X_Q. More generally, the spectral Galerkin approximation of the reaction-diffusion equation (Problem 2.22) consists of finding $\boldsymbol{u}_Q \in L^2([0,T]; X_Q)$ with

$$\begin{aligned} \left(\boldsymbol{u}_Q'(t), g_Q\right)_\Gamma + \mathfrak{h}(\boldsymbol{u}_Q(t), g_Q) &= \left(\mathcal{R}(\boldsymbol{u}_Q(t)), g_Q\right)_\Gamma \quad \text{for all} \quad g_Q \in X_Q,\ t > 0 \\ \boldsymbol{u}_Q(0) &= u_Q^0. \end{aligned} \tag{4.7}$$

4.2.1 Semidiscretized System

Choosing $\boldsymbol{u}_Q(0) = P_Q u^0$ and using the orthonormality of the basis functions $\phi \in \varphi_Q$, the spectral Galerkin approximation of the generalized heat equation (4.6) simplifies to the following semidiscretized system of differential equations.

Theorem 4.8 *Let Γ be a metric graph with ascending numbered eigenvalues $\lambda_q \in \sigma(\Gamma)$, $q = 1, \dots, Q$ and associated eigenfunctions $\phi_q \in \varphi_Q$. Let us further denote by*

$$\mathbf{c}(t) := (c_1(t), \dots, c_Q(t))^T$$

a vector collecting the coefficients of the expansion $\boldsymbol{u}_Q(t) = \sum_{q \leq Q} c_q(t)\phi_q$. Then, the spectral Galerkin approximation (4.6) *reduces to the initial value problem*

$$\begin{aligned} \frac{\mathrm{d}}{\mathrm{d}t}\mathbf{c}(t) + \boldsymbol{\Lambda}\mathbf{c}(t) &= \mathbf{f}(t) \\ \mathbf{c}(0) &= \mathbf{c}^0 \end{aligned} \tag{4.8}$$

where $\boldsymbol{\Lambda} := \mathrm{diag}(\{\lambda_q\}_{q \leq Q})$, the initial condition is given by

$$\mathbf{c}^0 := \left(\left(u^0, \phi_1\right)_\Gamma, \dots, \left(u^0, \phi_Q\right)_\Gamma\right)^T, \quad \textit{and}$$
$$\mathbf{f}(t) := \left((f(t), \phi_1)_\Gamma, \dots, (f(t), \phi_Q)_\Gamma\right)^T.$$

Proof Testing with an arbitrary but fixed $\phi_q \in \varphi_Q$ in (4.7) yields

$$\begin{aligned} \left(\boldsymbol{u}'_Q, \phi_q\right)_\Gamma + \mathfrak{h}\left(\boldsymbol{u}_Q, \phi_q\right) &= \frac{\mathrm{d}}{\mathrm{d}t}\left(\sum_{q' \leq Q} c_{q'}\phi_{q'}, \phi_q\right)_\Gamma + \mathfrak{h}\left(\sum_{q' \leq Q} c_{q'}\phi_{q'}, \phi_q\right) \\ &= \frac{\mathrm{d}}{\mathrm{d}t}c_q\left(\phi_q, \phi_q\right)_\Gamma + \sum_{q' \leq Q} c_{q'}\,\mathfrak{h}\left(\phi_{q'}, \phi_q\right) \end{aligned}$$

since $\left(\phi_{q'}, \phi_q\right)_\Gamma = 0$ for $\phi_{q'} \neq \phi_q$ (orthogonality). Thus, the spectral Galerkin approximation (4.7) can be written in matrix form as

$$\frac{\mathrm{d}}{\mathrm{d}t}\mathbf{M}\mathbf{c} + \mathbf{S}\mathbf{c} = \mathbf{f}(t)$$

where

$$\mathbf{M} := \begin{pmatrix} (\phi_1, \phi_1)_\Gamma & & 0 \\ & \ddots & \\ 0 & & (\phi_Q, \phi_Q)_\Gamma \end{pmatrix}, \quad \mathbf{S} := \begin{pmatrix} \mathfrak{h}(\phi_1, \phi_1) & \dots & \mathfrak{h}(\phi_Q, \phi_1) \\ \vdots & \ddots & \vdots \\ \mathfrak{h}(\phi_1, \phi_Q) & \dots & \mathfrak{h}(\phi_Q, \phi_Q) \end{pmatrix}.$$

Clearly, by orthonormality we obtain $(\phi_q, \phi_q)_\Gamma = 1$, i.e., $\mathbf{M} = \mathbf{I}$. Moreover, the eigenvalue identity in weak formulation implies $\mathfrak{h}(\phi_{q'}, \phi_q) = \lambda_{q'}(\phi_{q'}, \phi_q)_\Gamma$ and thereby also

$$\begin{aligned}\mathbf{S} = \begin{pmatrix} \mathfrak{h}(\phi_1, \phi_1) & \dots & \mathfrak{h}(\phi_Q, \phi_1) \\ \vdots & \ddots & \vdots \\ \mathfrak{h}(\phi_1, \phi_Q) & \dots & \mathfrak{h}(\phi_Q, \phi_Q) \end{pmatrix} &= \begin{pmatrix} \lambda_1(\phi_1, \phi_1)_\Gamma & \dots & \lambda_Q(\phi_Q, \phi_1)_\Gamma \\ \vdots & \ddots & \vdots \\ \lambda_1(\phi_1, \phi_Q)_\Gamma & \dots & \lambda_Q(\phi_Q, \phi_Q)_\Gamma \end{pmatrix} \\ &= \begin{pmatrix} \lambda_1(\phi_1, \phi_1)_\Gamma & 0 & 0 \\ 0 & \ddots & 0 \\ 0 & 0 & \lambda_Q(\phi_Q, \phi_Q)_\Gamma \end{pmatrix} \\ &= \mathbf{\Lambda}.\end{aligned}$$

□

Likewise, for the reaction-diffusion problem, we obtain the initial value problem

$$\begin{aligned}\frac{\mathrm{d}}{\mathrm{d}t}\mathbf{c}(t) + \mathbf{\Lambda}\mathbf{c}(t) &= \mathbf{r}(\mathbf{c}(t)) \\ \mathbf{c}(0) &= \mathbf{c}^0,\end{aligned} \tag{4.9}$$

with

$$\mathbf{r}(\mathbf{c}(t)) := \left((\mathcal{R}(\boldsymbol{u}_Q(t)), \phi_1)_\Gamma, \dots, (\mathcal{R}(\boldsymbol{u}_Q(t)), \phi_Q)_\Gamma\right)^T.$$

4.2.2 Error Analysis

Stability and convergence of the Galerkin approximation can be obtained by standard arguments, compare for example [12], Section 6.5.1. Yet, we are in the situation that the bilinear form $\mathfrak{h}$ is not coercive. However, we have seen that it fulfills Gårding's inequality (Sect. 2.3.2) and induces the seminorm $(\mathfrak{h}(u, u))^{\frac{1}{2}} = |u|_{H^1(\Gamma)}$. Therefore, a change of variable is required to obtain the desired results, see again [12], Section 6.5.1, Example 1.

For the generalized heat equation, the error of the spectral Galerkin discretization reduces to the truncation error of the exact solution.

Theorem 4.9 *Let $\boldsymbol{u}$ be the exact solution of the generalized heat equation* (2.15) *with $\boldsymbol{u}(t) \in \mathrm{dom}^d_{\mathcal{H},\mathrm{NK}}$ for $t > 0$. Then, the error of the spectral Galerkin approximation* (4.6) *on the finite dimensional subspace X_Q is given by*

$$\left\|\boldsymbol{u}_Q(t) - \boldsymbol{u}(t)\right\|_\Gamma \le \bar{C}\left(\frac{\mathrm{vol}_\Gamma}{(Q-n)\pi}\right)^{2d} \left\|\boldsymbol{u}(t)^{(2d)}\right\|_\Gamma$$

with $\bar{C}$ as in Corollary 4.7.

Proof A common technique is to write $\boldsymbol{u}_Q(t)-\boldsymbol{u}(t) = \boldsymbol{u}_Q(t)-P_Q\boldsymbol{u}(t)+P_Q\boldsymbol{u}(t)-\boldsymbol{u}(t)$ and estimate

$$\left\|\boldsymbol{u}_Q(t)-\boldsymbol{u}(t)\right\|_\Gamma \leq \left\|\boldsymbol{u}_Q(t)-P_Q\boldsymbol{u}(t)\right\|_\Gamma + \left\|P_Q\boldsymbol{u}(t)-\boldsymbol{u}(t)\right\|_\Gamma.$$

The second term is the truncation error of the solution at time t and for $\boldsymbol{u}(t) \in \mathrm{dom}^d_{\mathcal{H},\mathrm{NK}}$ can be bounded by

$$\left\|P_Q\boldsymbol{u}(t)-\boldsymbol{u}(t)\right\|_\Gamma \leq \bar{C}\left(\frac{\mathrm{vol}_\Gamma}{(Q-n)\pi}\right)^{2d}\left\|\boldsymbol{u}(t)^{(2d)}\right\|_\Gamma$$

(Corollaries 4.5 and 4.7). Let us now consider the error $\boldsymbol{e}(t) = \boldsymbol{u}_Q(t) - P_Q\boldsymbol{u}(t)$ and in the following notation omit the dependence on t to improve readability. For $g_Q \in X_Q$, the error satisfies

$$\begin{aligned}(\boldsymbol{e}', g_Q)_\Gamma + \mathfrak{h}(\boldsymbol{e}, g_Q) &= (\boldsymbol{u}'_Q, g_Q)_\Gamma + \mathfrak{h}(\boldsymbol{u}_Q, g_Q) - (P_Q\boldsymbol{u}', g_Q)_\Gamma - \mathfrak{h}(P_Q\boldsymbol{u}, g_Q)\\ &= (\boldsymbol{u}', g_Q)_\Gamma + \mathfrak{h}(\boldsymbol{u}, g_Q) - (P_Q\boldsymbol{u}', g_Q)_\Gamma - \mathfrak{h}(P_Q\boldsymbol{u}, g_Q)\\ &= (\boldsymbol{u}' - P_Q\boldsymbol{u}', g_Q)_\Gamma + \mathfrak{h}(\boldsymbol{u} - P_Q\boldsymbol{u}, g_Q)\end{aligned}$$

where the third equality follows from the fact that the spectral Galerkin solution $\boldsymbol{u}_Q$ fulfills (4.6). Since $g_Q \in X_Q$ is a linear combination of the eigenfunctions $\phi \in \varphi_Q$, their orthogonality implies the following:

1. $(\boldsymbol{u}' - P_Q\boldsymbol{u}', g_Q)_\Gamma = \left(\sum_{q>Q} c'_q\phi_q, g_Q\right)_\Gamma = 0.$
2. Since $\boldsymbol{u}(t) \in \mathrm{dom}_{\mathcal{H},\mathrm{NK}}$, we have

$$\begin{aligned}\mathfrak{h}(\boldsymbol{u} - P_Q\boldsymbol{u}, g_Q) = (\mathcal{H}(\boldsymbol{u} - P_Q\boldsymbol{u}), g_Q)_\Gamma &= \sum_{q>Q} c_q(\mathcal{H}\phi_q, g_Q)_\Gamma\\ &= \sum_{q>Q} c_q\lambda_q(\phi_q, g_Q)_\Gamma = 0.\end{aligned}$$

Here, we have used the eigenvalue identity and the commutability of the operators $\mathcal{H}$ and P_Q which follows from the fact that the coefficients in the expansion of $\mathcal{H}u$ are given by $(\mathcal{H}u, \phi_q)_\Gamma = \left(u, \phi''_q\right)_\Gamma = \lambda_q(u, \phi_q)_\Gamma$ for all q (this is only true for $u \in \mathrm{dom}_{\mathcal{H},\mathrm{NK}}$ as the self-adjointness is required in the second equality).

Choosing $g_Q = \boldsymbol{e}$ thus yields[3]

$$0 = \left(\boldsymbol{e}', g_Q\right)_\Gamma + \mathfrak{h}(\boldsymbol{e}, g_Q) = \frac{1}{2}\frac{\mathrm{d}}{\mathrm{d}t}\|\boldsymbol{e}\|_\Gamma^2 + \mathfrak{h}(\boldsymbol{e}, \boldsymbol{e}) \tag{4.10}$$

and, after integration,

$$0 = \frac{1}{2}\|\boldsymbol{e}(t)\|_\Gamma^2 - \|\boldsymbol{e}(0)\|_\Gamma^2 + \int_0^t \mathfrak{h}(\boldsymbol{e}(s), \boldsymbol{e}(s))\,\mathrm{d}s.$$

With the choice of the initial condition $\boldsymbol{u}_Q(0) = P_Q\boldsymbol{u}(0)$ we finally obtain $\|\boldsymbol{e}(0)\|_\Gamma = 0$, and, thus $\|\boldsymbol{e}(t)\|_\Gamma = 0$ which yields the assertion. □

In the semilinear case, the same arguments as in the first part of the proof lead to

$$\frac{1}{2}\frac{\mathrm{d}}{\mathrm{d}t}\|\boldsymbol{e}\|_\Gamma^2 + \mathfrak{h}(\boldsymbol{e}, \boldsymbol{e}) = \left(\mathcal{R}(\boldsymbol{u}) - \mathcal{R}(\boldsymbol{u}_Q), \boldsymbol{e}\right)_\Gamma$$

instead of (4.10). Using the Lipschitz continuity of $\mathcal{R}$ and Cauchy's inequality yields

$$\begin{aligned}\left(\mathcal{R}(\boldsymbol{u}) - \mathcal{R}(\boldsymbol{u}_Q), \boldsymbol{e}\right)_\Gamma &\leq C_L \left\|\boldsymbol{u} - \boldsymbol{u}_Q\right\|_\Gamma \left(\|\boldsymbol{e}\|_\Gamma + \sqrt{\mathfrak{h}(\boldsymbol{e}, \boldsymbol{e})}\right) \\ &\leq C_L \left(\left\|\boldsymbol{u} - P_Q\boldsymbol{u}\right\|_\Gamma + \left\|P_Q\boldsymbol{u} - \boldsymbol{u}_Q\right\|_\Gamma\right)\left(\|\boldsymbol{e}\|_\Gamma + \sqrt{\mathfrak{h}(\boldsymbol{e}, \boldsymbol{e})}\right) \\ &\leq C_L \left(\left\|\boldsymbol{u} - P_Q\boldsymbol{u}\right\|_\Gamma + \mathfrak{h}(\boldsymbol{e}, \boldsymbol{e}) + 2\|\boldsymbol{e}\|_\Gamma^2\right).\end{aligned}$$

Proceeding as in [50], proof of Theorem 13.1, we obtain after integration and applying Gronwall's lemma

$$\|\boldsymbol{e}\|_\Gamma^2 \leq \bar{C}_L \int_0^t \left\|\boldsymbol{u}(s) - P_Q\boldsymbol{u}(s)\right\|_\Gamma \mathrm{d}s.$$

Together with the truncation error, this yields

$$\begin{aligned}\left\|\boldsymbol{u}(t) - \boldsymbol{u}_Q(t)\right\|_\Gamma \leq \bar{C}\left(\frac{\mathrm{vol}_\Gamma}{(Q-n)\pi}\right)^{2d} \left\|\boldsymbol{u}(t)^{(2d)}\right\|_\Gamma \\ + \left(\bar{C}_L \int_0^t \left\|\boldsymbol{u}(s) - P_Q\boldsymbol{u}(s)\right\|_\Gamma \mathrm{d}s\right)^{\frac{1}{2}}\end{aligned}$$

with positive constants $\bar{C}, \bar{C}_L$.

[3] Note that $\frac{\mathrm{d}}{\mathrm{d}t}\|u\|_\Gamma^2 = 2(u, u)_\Gamma$ ([22], Theorem 3 in Section 5.9.2).

4.3 Solution of the Spectral Galerkin Semidiscretization

This section is devoted to the solution of the systems of ordinary differential equations arising from the spectral Galerkin discretization (4.8) and (4.9), respectively. For the generalized heat equation on Γ, the semidiscretized system reads

$$\frac{\mathrm{d}}{\mathrm{d}t}\mathbf{c}(t) + \boldsymbol{\Lambda}\mathbf{c}(t) = \mathbf{f}. \tag{4.11}$$

Multiplying (4.11) by $\exp(t\boldsymbol{\Lambda})$ leads to

$$\begin{aligned} &\exp(t\boldsymbol{\Lambda})\left(\frac{\mathrm{d}}{\mathrm{d}t}\mathbf{c}(t) + \boldsymbol{\Lambda}\mathbf{c}(t)\right) = \exp(t\boldsymbol{\Lambda})\,\mathbf{f} \\ \Leftrightarrow\quad &\frac{\mathrm{d}}{\mathrm{d}t}\left(\exp(t\boldsymbol{\Lambda})\mathbf{c}(t)\right) = \exp(t\boldsymbol{\Lambda})\,\mathbf{f}. \end{aligned}$$

Integrating both sides over $[0, s]$ yields

$$\begin{aligned} &\exp(s\boldsymbol{\Lambda})\mathbf{c}(s) = \mathbf{c}(0) + \int_0^s \exp(t\boldsymbol{\Lambda})\,\mathbf{f}\,\mathrm{d}t \\ \Leftrightarrow\quad &\mathbf{c}(s) = \exp(-s\boldsymbol{\Lambda})\mathbf{c}(0) + \int_0^s \exp\left((t-s)\boldsymbol{\Lambda}\right)\mathbf{f}\,\mathrm{d}t. \end{aligned} \tag{4.12}$$

Equation (4.12) is also known as the variation-of-constants formula. If $\mathbf{f}$ is constant over time, the exact solution at time s is readily given by

$$\mathbf{c}(s) = \exp(-s\boldsymbol{\Lambda})\mathbf{c}(s) + \begin{pmatrix} \Delta t & & & \\ & -\frac{\exp(-s\lambda_2)-1}{\lambda_2} & & \\ & & \ddots & \\ & & & -\frac{\exp(-s\lambda_n)-1}{\lambda_n} \end{pmatrix}\mathbf{f}.$$

4.3.1 *Exponential Integrators*

In the case of reaction-diffusion equations, the spectral Galerkin semidiscretization is given by

$$\frac{\mathrm{d}}{\mathrm{d}t}\mathbf{c}(t) + \boldsymbol{\Lambda}\mathbf{c}(t) = \mathbf{r}(\mathbf{c}(t)). \tag{4.13}$$

Here, the right hand side depends on $\mathbf{c}(t)$ and the integral in (4.12) cannot be computed in closed form. We therefore integrate over a small time interval $[0, \Delta t]$

only and obtain

$$\mathbf{c}(\Delta t) = \exp(-\Delta t\, \mathbf{\Lambda})\, \mathbf{c}(0) + \int_0^{\Delta t} \exp\left((t - \Delta t)\mathbf{\Lambda}\right) \mathbf{r}(\mathbf{c}(t))\, \mathrm{d}t. \tag{4.14}$$

The integral will now only be computed approximatively, for example as

$$\int_0^{\Delta t} \exp\left((t - \Delta t)\mathbf{\Lambda}\right) \mathbf{r}(\mathbf{c}(t))\, \mathrm{d}t \approx \int_0^{\Delta t} \exp\left((t - \Delta t)\mathbf{\Lambda}\right) \mathbf{r}(\mathbf{c}(0))\, \mathrm{d}t.$$

More generally, this idea can be extended to an iterative scheme to compute the coefficients at time $t + \Delta t$ by

$$\begin{aligned}\mathbf{c}(t + \Delta t) &= \exp(-\Delta t \mathbf{\Lambda})\, \mathbf{c}(t) + \int_0^{\Delta t} \exp((s - \Delta t)\mathbf{\Lambda})\ \mathbf{r}(\mathbf{c}(t))\mathrm{d}s \\ &= \exp(-\Delta t \mathbf{\Lambda})\, \mathbf{c}(t) + \begin{pmatrix} \Delta t & & & \\ & -\frac{\exp(-\Delta t \lambda_2)-1}{\lambda_2} & & \\ & & \ddots & \\ & & & -\frac{\exp(-\Delta t \lambda_n)-1}{\lambda_n} \end{pmatrix} \mathbf{r}(\mathbf{c}(t)).\end{aligned}$$

The presented approach is known as the exponential Euler method. Similarly, one can derive higher order exponential integrators if the integral is approximated by higher order Runge-Kutta methods, see for example [29]. Note that the advantage of exponential integrator methods in contrast to the IMEX approach used in Sect. 3.3 is the exact solution of the linear part of the differential equation. However, for the solution of finite element semidiscretizations, exponential integrators are much more expensive since the exponential of a non-diagonal matrix, or, to be more precise, the action of the exponential on a vector, has to be computed.

4.3.2 *Aspects of Implementation*

Computation of Inner Products

Besides the computation of an eigenfunction basis which will be treated in detail in the next section, the main computational costs of the spectral solution approach originate from the computation of inner products on graphs. To be more precise, we will address the computation of integrals over metric graphs involving the eigenfunctions ϕ_q, for example integrals of the form

$$\int_\Gamma \phi_q(x) f(x)\mathrm{d}x = \sum_{e \in \mathcal{E}} \int_e (\phi_q)_e(x) f_e(x)\mathrm{d}x$$

which are required to assemble the right hand side $\mathbf{f}$ in the semidiscretized generalized heat equation (4.8). In particular, the eigenfunctions ϕ_q on each edge are of the form

$$(\phi_q)_e(x) = A_e^q \cos\left(\sqrt{\lambda_q}\, x\right) + B_e^q \sin\left(\sqrt{\lambda_q}\, x\right)$$

as we will see later. For growing λ_q, the oscillation of ϕ_q increases such that classical numerical quadrature rules fail to provide reliable results. Instead, we will apply a method similar to *Filon's Quadrature* [23].

To increase readability, we will consider a fixed edge e and therefore suppress the edge subscripts in f, ϕ_q. The idea is to subdivide the interval $[0, \ell_e]$ in panels (subintervals) of length $2h_\kappa$ and consider the sub integrals

$$\int_e f(x)\phi_q(x)\mathrm{d}x = \int_0^{2h_\kappa} f(x)\phi_q(x)\mathrm{d}x + \ldots + \int_{\ell_e - 2h_\kappa}^{\ell_e} f(x)\phi_q(x)\mathrm{d}x.$$

On each panel $e_\kappa := [(\kappa-1)h_\kappa, (\kappa+1)h_\kappa]$, we then approximate f by

$$f(x) \approx \tilde{f}_\kappa(x) = \sum_{\iota=0}^{2} f(\xi_{\kappa,\iota})\psi_{\kappa,\iota}(x).$$

In this expression, $\psi_{\kappa,\iota}$ is a polynomial with maximal degree two and the quadrature nodes are $\xi_{\kappa,0} = (2\kappa-2)h_\kappa$, $\xi_{\kappa,1} = (2\kappa-1)h_\kappa$ and $\xi_{\kappa,2} = (2\kappa)h_\kappa$. The integrals $\int_{e_\kappa} \tilde{f}_\kappa(x)\phi_q(x)\mathrm{d}x$ can then be calculated in closed form by repeated partial integration.

To be more precise, let us consider the required integral on edge e given by

$$\begin{aligned}\int_e f(x)\phi_q \mathrm{d}x &= A_e^q \int_e f(x)\cos\left(\sqrt{\lambda_q}\, x\right)\mathrm{d}x + B_e^q \int_e f(x)\sin\left(\sqrt{\lambda_q}\, x\right)\mathrm{d}x \\ &=: A_e^q\, \mathcal{I}_e^{\cos}(\lambda_q) + B_e^q\, \mathcal{I}_e^{\sin}(\lambda_q)\end{aligned}$$

and focus on the integral in the left part of the sum, i.e., $\mathcal{I}_e^{\cos}(\lambda_q)$. As elaborated above, we approximate f on each panel e_κ by $\tilde{f}_\kappa$ and obtain

$$\mathcal{I}_e^{\cos}(\lambda_q) \approx \sum_{e_\kappa} \int_{e_\kappa} \tilde{f}_\kappa(x)\cos\left(\sqrt{\lambda_q}\, x\right)\mathrm{d}x$$

with

$$\begin{aligned}
\int_{e_\kappa} \tilde{f}_\kappa(x)\cos\left(\sqrt{\lambda_q}\,x\right)\mathrm{d}x &= \int_{e_\kappa}\sum_{\iota=0}^{2} f(\xi_{\kappa,\iota})\psi_{\kappa,\iota}(x)\cos\left(\sqrt{\lambda_q}\,x\right)\mathrm{d}x \\
&= \sum_{\iota=0}^{2} f(\xi_{\kappa,\iota})\int_{e_\kappa}\psi_{\kappa,\iota}(x)\cos\left(\sqrt{\lambda_q}\,x\right)\mathrm{d}x \\
&= \boldsymbol{m}_{e_\kappa}^{\cos}(\lambda_q)^T\mathcal{F}_{e_\kappa}
\end{aligned}$$

where

$$\mathcal{F}_{e_\kappa} := \left(f(\xi_{\kappa,1}), f(\xi_{\kappa,2}), f(\xi_{\kappa,3})\right)^T$$

and $\boldsymbol{m}_{e_\kappa}^{\cos}(\lambda_q)$ is a vector containing the integrals $\int_{e_\kappa}\psi_{\kappa,\iota}(x)\cos\left(\sqrt{\lambda_q}\,x\right)\mathrm{d}x$. Since each $\psi_{\kappa,\iota}$ is a polynomial of degree two, it can be expressed as

$$\psi_{\kappa,\iota}(x) = a_{\kappa,\iota}x^2 + b_{\kappa,\iota}x + c_{\kappa,\iota},$$

and thus

$$\begin{aligned}
&\int_{e_\kappa}\psi_{\kappa,\iota}(x)\cos\left(\sqrt{\lambda_q}\,x\right)\mathrm{d}x \\
&= \int_{e_\kappa}\left(a_{\kappa,\iota}x^2 + b_{\kappa,\iota}x + c_{\kappa,\iota}\right)\cos\left(\sqrt{\lambda_q}\,x\right)\mathrm{d}x \\
&= a_{\kappa,\iota}\int_{e_\kappa}x^2\cos\left(\sqrt{\lambda_q}\,x\right)\mathrm{d}x + b_{\kappa,\iota}\int_{e_\kappa}x\cos\left(\sqrt{\lambda_q}\,x\right)\mathrm{d}x + c_{\kappa,\iota}\int_{e_\kappa}\cos\left(\sqrt{\lambda_q}\,x\right)\mathrm{d}x \\
&=: a_{\kappa,\iota}\,\mathcal{I}_\kappa^{\cos,\,x^2}(\lambda_q) + b_{\kappa,\iota}\,\mathcal{I}_\kappa^{\cos,\,x}(\lambda_q) + c_{\kappa,\iota}\,\mathcal{I}_\kappa^{\cos}(\lambda_q).
\end{aligned}$$

The remaining integrals can be evaluated using their closed forms given in Table 4.1. The vector $\boldsymbol{m}_{e_\kappa}^{\cos}(\lambda_q)$ can consequently be assembled as

$$\boldsymbol{m}_{e_\kappa}^{\cos}(\lambda_q) = \begin{pmatrix} a_{\kappa,1} & b_{\kappa,1} & c_{\kappa,1} \\ a_{\kappa,2} & b_{\kappa,2} & c_{\kappa,2} \\ a_{\kappa,3} & b_{\kappa,3} & c_{\kappa,3} \end{pmatrix}\begin{pmatrix} \mathcal{I}_\kappa^{\cos,\,x^2}(\lambda_q) \\ \mathcal{I}_\kappa^{\cos,\,x}(\lambda_q) \\ \mathcal{I}_\kappa^{\cos}(\lambda_q) \end{pmatrix}.$$

The integral $\mathcal{I}_e^{\cos}(\lambda_q)$ on the whole edge is then given by

$$\mathbf{1}^T\left(\boldsymbol{m}_e^{\cos}(\lambda_q)\,\mathcal{F}_e\right)$$

where

$$\mathcal{F}_e = (f(h_\kappa), f(2h_\kappa), \dots, f(N_\kappa h_\kappa))^T$$

and

$$\boldsymbol{m}_e^{\cos}(\lambda_q) = \begin{pmatrix} \boxed{\boldsymbol{m}_{e_1}^{\cos}(\lambda_q)^T} & 0\;0 & \dots & & & & \\ 0\;0 & \boxed{\boldsymbol{m}_{e_2}^{\cos}(\lambda_q)^T} & 0\;\;0 & \dots & & & \\ \dots 0 & & 0 & \boxed{\boldsymbol{m}_{e_3}^{\cos}(\lambda_q)^T} & 0\;\;0 \dots & & \\ & & & \ddots & \ddots & & \\ & & & 0\;\;0 & \boxed{\boldsymbol{m}_{e_{N_\kappa-1}}^{\cos}(\lambda_q)^T} & 0\;0 & \\ & & & & 0\;\;0 & \boxed{\boldsymbol{m}_{e_{N_\kappa-1}}^{\cos}(\lambda_q)^T} & \end{pmatrix}.$$

Clearly, the same considerations apply for the second integral $\mathcal{I}_e^{\sin}(\lambda_q)$, for which we also summarized the closed form integrals $\mathcal{I}_\kappa^{\sin, x^2}(\lambda_q), \mathcal{I}_\kappa^{\sin, x}(\lambda_q), \mathcal{I}_\kappa^{\sin}(\lambda_q)$ in Table 4.1. Together, we have

$$\int_e f(x)\phi_q(x)\mathrm{d}x \approx A_e^q\, \mathbf{1}^T \left(\boldsymbol{m}_e^{\cos}(\lambda_q)\mathcal{F}_e\right) + B_e^q\, \mathbf{1}^T \left(\boldsymbol{m}_e^{\sin}(\lambda_q)\mathcal{F}_e\right).$$

It is worthwhile mentioning that $\boldsymbol{m}_e^{\cos}$ and $\boldsymbol{m}_e^{\sin}$ do not depend on f and therefore can be computed and stored in advance to be reused for different f. This is in particular important when we return to the general case where $\mathcal{R}$ depends on u and therefore $\left(\mathcal{R}(u_Q(t)), \phi_q\right)_\Gamma$ is not constant over time. Moreover, for equilateral graphs, $\boldsymbol{m}_e^{\cos}$ and $\boldsymbol{m}_e^{\sin}$ are identical for all edges and only $\mathcal{F}_e$ as well as the constants A_e^q, B_e^q need to be modified.

Evaluation of the Spectral Expansion

In the nonlinear case, i.e., towards the computation of integrals of the form

$$\left(\mathcal{R}(\boldsymbol{u}_Q(t)), \phi_q\right)_\Gamma = \sum_{e\in\mathcal{E}} \int_e \mathcal{R}((\boldsymbol{u}_Q)_e(t))(\phi_q)_e(x)\mathrm{d}x,$$

it moreover arises the question of efficiently evaluating the spectral expansion $\boldsymbol{u}_Q$. In particular, at the nodes of the quadrature formula derived above, we have to assemble the equivalent of the vector $\mathcal{F}_e$, now given by

$$\boldsymbol{r}_e := (\mathcal{R}((u_Q)_e(h_\kappa, t)), \mathcal{R}((u_Q)_e(2h_\kappa, t)), \dots, \mathcal{R}((u_Q)_e(N_\kappa, t)))$$

Table 4.1 Closed form integrals for Filon quadrature

$\mathcal{I}_{\kappa}^{\cos, x^2}(\lambda_q)$	$\frac{1}{\lambda_q^{3/2}}\Big(-(h_\kappa^2(\kappa-1)^2\lambda_q - 2)\sin\big(h_\kappa(\kappa-1)\sqrt{\lambda_q}\big) + (h_\kappa^2(\kappa+1)^2\lambda_q - 2)\sin\big(h_\kappa(\kappa+1)\sqrt{\lambda_q}\big) - 2h_\kappa(\kappa-1)\sqrt{\lambda_q}\cos\big(h_\kappa(\kappa-1)\sqrt{\lambda_q}\big) + 2h_\kappa(\kappa+1)\sqrt{\lambda_q}\cos\big(h_\kappa(\kappa+1)\sqrt{\lambda_q}\big)\Big)$
$\mathcal{I}_{\kappa}^{\cos, x}(\lambda_q)$	$\frac{2}{\lambda_q}\Big(h_\kappa\kappa\sqrt{\lambda_q}\sin\big(h_\kappa\sqrt{\lambda_q}\big)\cos\big(h_\kappa\kappa\sqrt{\lambda_q}\big) + \sin\big(h_\kappa\kappa\sqrt{\lambda_q}\big)\Big(h_\kappa\sqrt{\lambda_q}\cos\big(h_\kappa\sqrt{\lambda_q}\big) - \sin\big(h_\kappa\sqrt{\lambda_q}\big)\Big)\Big)$
$\mathcal{I}_{\kappa}^{\cos}(\lambda_q)$	$\frac{2}{\sqrt{\lambda_q}}\Big(\sin\big(h_\kappa\sqrt{\lambda_q}\big)\cos\big(h_\kappa\kappa\sqrt{\lambda_q}\big)\Big)$
$\mathcal{I}_{\kappa}^{\sin, x^2}(\lambda_q)$	$\frac{1}{\lambda_q^{3/2}}\Big((h_\kappa^2(\kappa-1)^2\lambda_q - 2)\cos\big(h_\kappa(\kappa-1)\sqrt{\lambda_q}\big) - (h_\kappa^2(\kappa+1)^2\lambda_q - 2)\cos\big(h_\kappa(\kappa+1)\sqrt{\lambda_q}\big) - 2h_\kappa(\kappa-1)\sqrt{\lambda_q}\sin\big(h_\kappa(\kappa-1)\sqrt{\lambda_q}\big) + 2h_\kappa(\kappa+1)\sqrt{\lambda_q}\sin\big(h_\kappa(\kappa+1)\sqrt{\lambda_q}\big)\Big)$
$\mathcal{I}_{\kappa}^{\sin, x}(\lambda_q)$	$\frac{2}{\lambda_q}\Big(\sin\big(h_\kappa\sqrt{\lambda_q}\big)\Big(h_\kappa\kappa\sqrt{\lambda_q}\sin\big(h_\kappa\kappa\sqrt{\lambda_q}\big) + \cos\big(h_\kappa\kappa\sqrt{\lambda_q}\big)\Big) - h_\kappa\sqrt{\lambda_q}\cos\big(h_\kappa\sqrt{\lambda_q}\big)\cos\big(h_\kappa\kappa\sqrt{\lambda_q}\big)\Big)$
$\mathcal{I}_{\kappa}^{\sin}(\lambda_q)$	$\frac{2}{\sqrt{\lambda_q}}\Big(\sin\big(h_\kappa\sqrt{\lambda_q}\big)\sin\big(h_\kappa\kappa\sqrt{\lambda_q}\big)\Big)$

on each edge. Again, to increase readability, we will suppress the subscript in $(\boldsymbol{u}_Q)_e$ which should not lead to any confusion since we will once more start to consider a fixed edge e. On this edge, the spectral expansion of $\boldsymbol{u}$, evaluated at a node κh_κ, is given by

$$u_Q(\kappa h_\kappa, t) = \sum_{q=1}^{Q} c_q(t)\phi_q(\kappa h_\kappa) = (\phi_1(\kappa h_\kappa), \ldots, \phi_Q(\kappa h_\kappa))\,\mathbf{c}(t).$$

Obviously, the evaluation of the eigenfunctions does not depend on t and thus can be computed in advance as

$$\phi_q(\kappa h_\kappa) = A_e^q \cos\left(\sqrt{\lambda_q}\,\kappa h_\kappa\right) + B_e^q \sin\left(\sqrt{\lambda_q}\,\kappa h_\kappa\right).$$

Defining

$$\boldsymbol{\phi}_\kappa^{\cos} := \left(\cos\left(\sqrt{\lambda_1}\,\kappa h_\kappa\right), \ldots, \cos\left(\sqrt{\lambda_Q}\,\kappa h_\kappa\right)\right)$$

and $\boldsymbol{\phi}_\kappa^{\sin}$ equivalently yields

$$u_Q(\kappa h_\kappa, t) = (\boldsymbol{\phi}_\kappa^{\cos}\mathbf{A}_e + \boldsymbol{\phi}_\kappa^{\sin}\mathbf{B}_e)\mathbf{c}(t)$$

where $\mathbf{A}_e := \operatorname{diag}(A_e^1, \ldots, A_e^Q)$ and $\mathbf{B}_e := \operatorname{diag}(B_e^1, \ldots, B_e^Q)$. The evaluations of $\boldsymbol{u}_Q$ on all nodes can consequently be obtained by

$$\begin{aligned}(u_Q(h_\kappa, t), \ldots, u_Q(N_\kappa, t))^T &= ((\boldsymbol{\phi}_1^{\cos}, \ldots, \boldsymbol{\phi}_{N_\kappa}^{\cos})\mathbf{A}_e + (\boldsymbol{\phi}_1^{\sin}, \ldots, \boldsymbol{\phi}_{N_\kappa}^{\sin})\mathbf{B}_e)\,\mathbf{c}(t)\\ &=: (\boldsymbol{\phi}_e^{\cos}\mathbf{A}_e + \boldsymbol{\phi}_e^{\sin}\mathbf{B}_e)\,\mathbf{c}(t).\end{aligned}$$

All the coefficients in $\boldsymbol{\phi}_e^{\cos}$ and $\boldsymbol{\phi}_e^{\sin}$ can be computed in advance and, moreover, in the special case of equilateral graphs, they do not depend on the edge and thus only need to be computed once for all edges. However, also in the case of non-equilateral graphs, a careful choice of the interpolation nodes can guarantee the reuse of the function evaluation $\cos(\sqrt{\lambda_q}\,\kappa h_\kappa)$ and $\sin(\sqrt{\lambda_q}\,\kappa h_\kappa)$ across the edges.

4.4 Application to Other Classes of Partial Differential Equations

With our application to the simulation of protein distribution in mind, we have so far focused our work on classical diffusion-type equations. However, we would like to emphasize that the derived spectral method can be readily applied to other PDEs involving the operator $\mathcal{H}$ from (2.1). For instance, consider a wave equation on a metric graph given by

$$\frac{\partial^2}{\partial t^2}u + \mathcal{H}u = 0 \tag{4.15}$$

subject to Neumann-Kirchhoff boundary conditions and along with initial conditions $u(0) = u^0$ and $\frac{\partial}{\partial t}u(0) = u^{0,t}$. We can use the exact same argumentation as for the semilinear parabolic equation to derive the Galerkin approximation:

$$\left(\boldsymbol{u}_Q''(t), \phi\right)_\Gamma + \mathfrak{h}(\boldsymbol{u}_Q(t), \phi) = 0 \quad \text{for all} \quad \phi \in \varphi_Q, t > 0$$

$$\boldsymbol{u}_Q(0) = u_Q^0, \ \boldsymbol{u}_Q'(0) = u_Q^{0,t}$$

which reduces to an ODE of the form

$$\frac{\mathrm{d}^2}{\mathrm{d}t^2}\mathbf{c}(t) + \boldsymbol{\Lambda}\mathbf{c}(t) = 0$$

with $\mathbf{c}$ as in Theorem 4.8.

Remarkably, the same applies to fractional diffusion equations of the type

$$\frac{\partial}{\partial t}u - \Delta^\alpha u = 0$$

with $\alpha \in \mathbb{R}$. To see this, we make us of the spectral representation of the fractional Laplacian given by

$$\Delta^\alpha u = \sum_\lambda \lambda^\alpha c_\lambda \phi_\lambda.$$

Consequently, if we define $\boldsymbol{\Lambda}^\alpha = \operatorname{diag}\left(\lambda_1^\alpha, \ldots, \lambda_Q^\alpha\right)$, we can simply replace $\boldsymbol{\Lambda}$ by $\boldsymbol{\Lambda}^\alpha$ in (4.8) to deduce the semidiscretization

$$\frac{\mathrm{d}}{\mathrm{d}t}\mathbf{c}(t) + \boldsymbol{\Lambda}^\alpha\mathbf{c}(t) = 0.$$

Chapter 5
Computation of Quantum Graph Spectra

Abstract The spectrum of a quantum graph is defined as the spectrum of the negative second order derivative with Neumann-Kirchhoff boundary conditions, which consists exclusively of its eigenvalues. Speaking of the computation of quantum graph spectra, this always involves the associated eigensolutions of the eigenvalue problem, referred to as eigenfunctions.

In this chapter, an algorithm to compute an arbitrarily large part of the lower spectrum of a quantum graph along with a representation of the corresponding eigenfunctions in closed form is proposed. The theoretical foundation is a very remarkable relation of the continuous problem to a nonlinear eigenvalue problem or, in the special case of equilateral graphs, to a linear eigenvalue problem, both only determined on the vertices of the underlying combinatorial graph. Yet this finding does not apply to a particular part of the spectrum, which will be refer to as *non-vertex spectrum*. Section 5.2 then derives how this relation, nevertheless, can be applied to develop a practical algorithm in the special case of equilateral graphs and for both parts of the spectrum. A generalization to non-equilateral graphs is discussed in Sect. 5.3.

New aspects of this chapter are the exploitation of the prominent relation to combinatorial graphs for the whole spectrum using an extended graph, the arising numerical algorithm as well as the application of the latter to solve the nonlinear eigenvalue problem (NEP) in the non-equilateral case.

5.1 Relation to Combinatorial Graph Spectra

We briefly recall that, according to Theorem 4.2, $\lambda \in \mathbb{R}^+$ is an eigenvalue of Γ if there exists a nontrivial $\phi \in \mathrm{dom}_{\mathcal{H},\mathrm{NK}}$ with

$$\mathcal{H}\phi = \lambda\phi. \tag{5.1}$$

In particular, the restriction of ϕ to the vertices of Γ (denoted by $\phi_{\mathcal{V}}$) is well defined since ϕ is continuous.

A. Weller, *Numerical Methods for Metric Graphs*, Lecture Notes in Mathematics 2382, https://doi.org/10.1007/978-3-032-05011-3_5

An important result for the following numerical computations is the relation of the continuous eigenvalue problem (5.1) to a discrete eigenvalue problem in terms of graph matrices associated with the underlying combinatorial graph $\mathcal{G}$. As pointed out in [6], such a relation cannot be assumed immediately since the spectrum of $\mathcal{H}$ is unbounded (in contrast to the spectrum of a discrete matrix). Surprisingly, it turns out that, especially for equilateral graphs, a very useful relation exists and has been discovered by several authors in various contexts and variants, see, for example, [6, 13, 42, 44, 53], just to mention a few. In all of these investigations as well as in this book, this special relation only holds for a specific part of the spectrum which we decided to refer to as the *vertex spectrum*.

Definition 5.1 The spectrum of a compact, finite metric graph Γ with edge lengths $\ell_e, e \in \mathcal{E}$, can be classified into the *vertex spectrum*

$$\sigma_{\mathcal{V}}(\Gamma) := \left\{\lambda \in \sigma(\Gamma) : \lambda \neq \left(\frac{k\pi}{\ell_e}\right)^2 \text{ for all } k \in \mathbb{N}_0 \text{ and } \ell_e, e \in \mathcal{E}\right\}$$

and the *non-vertex spectrum*

$$\sigma_{\mathcal{E}}(\Gamma) := \left\{\lambda \in \sigma(\Gamma) : \lambda = \left(\frac{k\pi}{\ell_e}\right)^2 \text{ for } k \in \mathbb{N}_0 \text{ and } \ell_e, e \in \mathcal{E}\right\}.$$

I would like to point out that in some of the works cited above, the non-vertex spectrum is referred to as *Dirichlet-Spectrum*. This is motivated by the fact that these eigenvalues occur if the associated eigenfunction fulfills some sort of Dirichlet condition at the vertices. However, my impression is that not much attention is paid to the non-vertex spectrum, especially not to the behavior of the associated eigenfunctions across the edges. Therefore, we will carefully treat the non-vertex part of the spectrum and propose a novel method to compute the associated eigenfunctions in the following subsection. Yet, for the remainder of this subsection, we focus on vertex eigenvalues and their relation to discrete graph matrices.

We first show that the vertex eigenvalues of a non-equilateral graph can be found as the solution of a so-called *Nonlinear Eigenvalue Problem* (NEP). The remarkable aspect of this NEP is that it is a system of size $n \times n$ and its solution vector is only defined at the vertices of Γ. This motivated us to choose the name *vertex eigenvalues*.

Theorem 5.2 *Let Γ be a compact metric graph. Then, $\lambda \in \sigma_{\mathcal{V}}(\Gamma)$ if and only if there exists a nontrivial $\Phi \in \mathbb{R}^n$ such that*

$$\mathbf{H}(\lambda)\Phi = 0$$

where the matrix $\mathbf{H}(\lambda) \in \mathbb{R}^{n \times n}$ *is given by*

$$\mathbf{H}_{ij}(\lambda) := \begin{cases} -\sum_{e \in \mathcal{E}_{v_i}} \cot\left(\sqrt{\lambda}\,\ell_e\right) & \text{if } i = j \\ \frac{1}{\sin\left(\sqrt{\lambda}\,\ell_e\right)} & \text{if } e = (v_i, v_j) \in \mathcal{E} \\ 0 & \text{otherwise.} \end{cases} \tag{5.2}$$

Proof The main part of the proof is conducted according to the observations given in [6], Chapter 3.6.

In general, the solution of the eigenvalue problem (5.1) on each edge $e = (v_i, v_j)$ has the form

$$\phi_e(x) = A_e \cos\left(\sqrt{\lambda} x\right) + B_e \sin\left(\sqrt{\lambda} x\right), \quad x \in [0, \ell_e] \tag{5.3}$$

for $\lambda > 0$ and constants $A_e, B_e \in \mathbb{R}$. The eigenfunction ϕ on the whole metric graph is then given by the collection $\{\phi_e\}_{e \in \mathcal{E}}$.

In particular, the restriction $\Phi := \phi_{\mathcal{V}} \in \mathbb{R}^n$ of ϕ to the vertices $v \in \mathcal{V}$ is well defined since the eigenfunctions ϕ must be continuous on Γ. This implies that the boundary values of (5.3) are given by $\phi_e(0) = \Phi(v_i)$ and $\phi_e(\ell_e) = \Phi(v_j)$ (recall that $e = (v_i, v_j)$, i.e., $o(e) = v_i$ and $t(e) = v_j$). We may than reformulate (5.3) to determine the constants A_e and B_e in dependence of the boundary values and obtain

$$\Phi(v_i) = \phi_e(0) = A_e$$

and

$$\Phi(v_j) = \phi_e(\ell_e) = \Phi(v_i) \cos\left(\sqrt{\lambda}\,\ell_e\right) + B_e \sin\left(\sqrt{\lambda}\,\ell_e\right),$$

i.e.,

$$B_e = \frac{1}{\sin(\sqrt{\lambda}\,\ell_e)} \left(\Phi(v_j) - \Phi(v_i) \cos\left(\sqrt{\lambda}\,\ell_e\right)\right).$$

Note that we excluded possible eigenvalues of the form $\lambda = \left(\frac{k\pi}{\ell_e}\right)^2$ for $k \in \mathbb{N}_0$ and $\ell_e, e \in \mathcal{E}$ guaranteeing that $\sin(\sqrt{\lambda}\ell_e) \neq 0$. Substituting the constants finally yields that the eigenfunctions on the edges can be expressed in terms of the boundary conditions as

$$\begin{aligned}
\phi_e(x) &= \Phi(v_i)\cos(\sqrt{\lambda}x) + \frac{1}{\sin(\sqrt{\lambda}\,\ell_e)}\Big(\Phi(v_j) - \Phi(v_i)\cos(\sqrt{\lambda}\,\ell_e)\Big)\sin(\sqrt{\lambda}x)\\
&= \frac{1}{\sin(\sqrt{\lambda}\,\ell_e)}\Big(\Phi(v_i)\cos(\sqrt{\lambda}x)\sin(\sqrt{\lambda}\,\ell_e)\\
&\quad + \Big(\Phi(v_j) - \Phi(v_i)\cos(\sqrt{\lambda}\,\ell_e)\Big)\sin(\sqrt{\lambda}x)\Big)\\
&= \frac{1}{\sin(\sqrt{\lambda}\,\ell_e)}\Big(\Phi(v_i)\Big(\cos(\sqrt{\lambda}x)\sin(\sqrt{\lambda}\,\ell_e)\\
&\quad - \cos(\sqrt{\lambda}\,\ell_e)\sin(\sqrt{\lambda}x)\Big) + \Phi(v_j)\sin(\sqrt{\lambda}x)\Big)\\
&= \frac{1}{\sin(\sqrt{\lambda}\,\ell_e)}\Big(\Phi(v_i)\sin(\sqrt{\lambda}(\ell_e - x)) + \Phi(v_j)\sin(\sqrt{\lambda}x)\Big)\,. \qquad (5.4)
\end{aligned}$$

Besides the continuity condition we used to reformulate each ϕ_e in terms of the boundary values $\Phi(v_i)$ and $\Phi(v_j)$, we further need to guarantee that ϕ satisfies the conservation of currents condition $\mathcal{K}(v) = 0$[1] for all $v \in \mathcal{V}$. Therefore, we consider the derivative

$$\phi_e'(x) = \frac{\sqrt{\lambda}}{\sin(\sqrt{\lambda}\,\ell_e)}\Big(\cos\Big(\sqrt{\lambda}x\Big)\,\Phi(v_j) - \cos\Big(\sqrt{\lambda}\,(\ell_e - x)\Big)\,\Phi(v_i)\Big)$$

and demand

$$\sum_{e\in\mathcal{E}_{v_i}} \phi_e'(v_i) = \sum_{v_j\sim v_i} \frac{\sqrt{\lambda}}{\sin(\sqrt{\lambda}\,\ell_e)}\Big(\Phi(v_j) - \cos\Big(\sqrt{\lambda}\,\ell_e\Big)\,\Phi(v_i)\Big) = 0 \qquad (5.5)$$

for all $v_i \in \mathcal{V}$. This is a necessary and sufficient condition for ϕ to be a nontrivial solution of the eigenvalue problem and it is equivalent to

$$\sum_{v_j\sim v_i} \frac{1}{\sin(\sqrt{\lambda}\,\ell_e)}\Phi(v_j) - \sum_{v_j\sim v_i} \frac{\cos(\sqrt{\lambda}\,\ell_e)}{\sin(\sqrt{\lambda}\,\ell_e)}\Phi(v_i) = 0$$

for all $v \in \mathcal{V}$, or, in matrix form

$$\mathbf{H}(\lambda)\Phi = 0$$

with **H** as given in (5.2). □

[1] $(\mathcal{K}u)(v) := \sum_{e\in\mathcal{E}_v} \frac{\mathrm{d}u_e}{\mathrm{d}x}(v)$, as defined in (2.4).

The fact that the eigenvalue problem on Γ can be reduced to an NEP as in Theorem 5.2 was also observed in [35], although no explicit characterization of $\mathbf{H}$ is given and the NEP is not further investigated.

In the special case of an equilateral graph Γ, an even stronger relationship can be identified.

Theorem 5.3 *Let Γ be a compact, equilateral metric graph with edge length ℓ and underlying combinatorial graph $\mathcal{G}$. Then, $\lambda \in \sigma_{\mathcal{V}}(\Gamma)$ if and only if*

$$\left(1 - \cos\left(\sqrt{\lambda}\ell\right)\right) \in \sigma(\Delta_{\mathcal{G}})\backslash\{0, 2\}$$

where $\Delta_{\mathcal{G}} = \mathbf{D}^{-1}\mathbf{L}$ is the harmonic graph Laplacian of the underlying combinatorial graph $\mathcal{G}$.

Proof The harmonic graph Laplacian acting on a discrete function $\mathrm{u} \in \mathbb{R}^n$ on $\mathcal{G}$ takes the form

$$\Delta_{\mathcal{G}}\mathrm{u}(v_i) = \mathrm{u}(v_i) - \frac{1}{\deg(v_i)} \sum_{v_j \sim v_i} \mathrm{u}(v_j).$$

Thus, setting $\ell_e = \ell$ for all edges, we can simplify (5.5) in the proof of Theorem 5.2 to

$$\sum_{e \in \mathcal{E}_{v_i}} \phi_e'(v_i) = \sum_{v_j \sim v_i} \frac{\sqrt{\lambda}}{\sin(\sqrt{\lambda}\,\ell)} \left(\Phi(v_j) - \cos(\sqrt{\lambda}\,\ell)\Phi(v_i)\right) = 0$$

if and only if

$$\sum_{v_j \sim v_i} \left(\Phi(v_j) - \cos(\sqrt{\lambda}\,\ell)\Phi(v_i)\right) = \sum_{v_j \sim v_i} \Phi(v_j) - \cos(\sqrt{\lambda}\,\ell)\deg(v_i)\Phi(v_i) = 0$$

which is equivalent to

$$\frac{1}{\deg(v_i)} \sum_{v_j \sim v_i} \Phi(v_j) = \cos(\sqrt{\lambda}\,\ell)\Phi(v_i).$$

Adding $\Phi(v_i)$ on both sides yields

$$\Phi(v_i) - \frac{1}{\deg(v_i)} \sum_{v_j \sim v_i} \Phi(v_j) = \Phi(v_i) - \cos(\sqrt{\lambda}\,\ell)\Phi(v_i),$$

what we can finally rewrite as the eigenvalue equation $\Delta_{\mathcal{G}}\Phi(v_i) = \left(1 - \cos\left(\sqrt{\lambda}\,\ell\right)\right)\Phi(v_i)$ for all $v_i \in \mathcal{V}$. □

I once more emphasize that Theorem 5.3 only applies if $\sin\left(\sqrt{\lambda}\,\ell\right) \neq 0$ since this assured the representation of ϕ_e in terms of the boundary values to be well defined. We will carefully treat the excluded *non-vertex* part of the spectrum in Sect. 5.2.2.

5.2 Spectra of Equilateral Graphs

In this section, the computation of the vertex and non-vertex spectrum based on Theorem 5.3 is studied for equilateral quantum graphs. The presentation is accompanied by two minimal examples to illustrate the described method and derived patterns in the spectrum of metric graphs. The results of this section have been prepared in joint work with Prof. Dr. Mark Ainsworth and are partly published in [1]. However, several aspects have been elaborated with additional details and more examples for this work.

5.2.1 Vertex Spectrum

In Theorem 5.3, we have only concentrated on the eigenvalues and somewhat neglected the eigenfunctions. From the proof of Theorem 5.3, however, it directly follows that the restriction of ϕ to the vertices is given by the eigenvectors Φ of $\Delta_{\mathcal{G}}$. Slightly rephrased, we would therefore like to state the following theorem.

Theorem 5.4 *Let* $\sin\left(\sqrt{\lambda}\,\ell\right) \neq 0$. *Then,* λ *is an eigenvalue of* Γ *corresponding to the eigenfunction* ϕ *defined by*

$$\phi_e = \frac{1}{\sin\left(\sqrt{\lambda}\,\ell\right)} \left(\Phi(v_i)\sin\left(\sqrt{\lambda}(\ell - x)\right) + \Phi(v_j)\sin\left(\sqrt{\lambda}x\right)\right)$$

if and only if $\mu = \left(1 - \cos\left(\sqrt{\lambda}\,\ell\right)\right)$ *is an eigenvalue of* $\Delta_{\mathcal{G}}$ *corresponding to the eigenvector* $\Phi = (\Phi(v_1), \ldots, \Phi(v_n))^T$.

Theorem 5.4 provides us with a simple procedure to compute an arbitrarily large part of the lower vertex spectrum using only eigenvalues and eigenvectors of the discrete $n \times n$ matrix $\Delta_{\mathcal{G}}$.

Corollary 5.5 *Any vertex eigenvalue* λ *and associated vertex eigenfunction* ϕ *can be determined by the rule*

$$\lambda_{\mu,k} = \begin{cases} \left(\frac{1}{\ell}\left(\arccos(1-\mu) + k\pi\right)\right)^2 & \text{for } k \text{ even} \\ \left(\frac{1}{\ell}\left(-\arccos(1-\mu) + (k+1)\pi\right)\right)^2 & \text{for } k \text{ odd} \end{cases} \tag{5.6}$$

Table 5.1 Bunches of vertex eigenvalues in increasing order

μ	k	$\lambda_{\mu,k}$	
	$k=0$	$\lambda_{\mu,0} = \left(\frac{1}{\ell}(\arccos(1-\mu))\right)^2$	$\in \left(0, (\frac{\pi}{\ell})^2\right)$
	$k=1$	$\lambda_{\mu,1} = \left(\frac{1}{\ell}(-\arccos(1-\mu)+2\pi)\right)^2$	$\in \left((\frac{\pi}{\ell})^2, (\frac{2\pi}{\ell})^2\right)$
$\mu \in (0,2)$	$k=2$	$\lambda_{\mu,2} = \left(\frac{1}{\ell}(\arccos(1-\mu)+2\pi)\right)^2$	$\in \left((\frac{2\pi}{\ell})^2, (\frac{3\pi}{\ell})^2\right)$
	$\vdots$	$\vdots$	

and

$$(\phi_{\mu,k})_e(x) = \frac{1}{\sin\left(\sqrt{\lambda_{\mu,k}}\,\ell\right)}\left(\Phi(v_i)\sin\left(\sqrt{\lambda_{\mu,k}}\,(\ell-x)\right)+\Phi(v_j)\sin\left(\sqrt{\lambda_{\mu,k}}\,x\right)\right) \tag{5.7}$$

where (μ, Φ) is an eigenpair of $\Delta_{\mathcal{G}}$ with $\mu \notin \{0,2\}$ and $k \in \mathbb{N}_0$.

Proof Note that in order to avoid $\sin\left(\sqrt{\lambda_{\mu,k}}\,\ell\right) = 0$, we need to exclude possible eigenvalues $\mu \in \{0,2\}$ of $\Delta_{\mathcal{G}}$. The assertion then directly follows from Theorem 5.4 and rearranging by $\lambda_{\mu,k}$. □

In fact, as the eigenvalues of $\Delta_{\mathcal{G}}$ lie within the interval $[0,2]$, rule (5.6) naturally provides us with bunches of eigenvalues coming in increasing order, compare Table 5.1. In general, $\lambda_{\mu,k} \in \sigma_{\mathcal{V}}(\Gamma)$ lies in the interval $\left(\left(\frac{k\pi}{\ell}\right)^2, \left(\frac{(k+1)\pi}{\ell}\right)^2\right)$. This motivates the following definition.

Definition 5.6 The *k-th bunch of vertex eigenvalues* is given by

$$\sigma_{\mathcal{V},k}(\Gamma) := \{\lambda_{\mu,k} : \mu \in \sigma(\Delta_{\mathcal{G}})\setminus\{0,2\}\}$$

with

$$\sigma_{\mathcal{V},k}(\Gamma) \subseteq \left(\left(\frac{k\pi}{\ell}\right)^2, \left(\frac{(k+1)\pi}{\ell}\right)^2\right).$$

We will later be especially interested in $\sigma_{\mathcal{V},0}(\Gamma)$ which contains all eigenvalues $0 < \lambda < \left(\frac{\pi}{\ell}\right)^2$. Finally, we observe that the cardinality of each bunch can be determined a priori.

Corollary 5.7 *For each $k \in \mathbb{N}_0$, the cardinality of $\sigma_{\mathcal{V},k}(\Gamma)$ is given by*

$$|\sigma_{\mathcal{V},k}(\Gamma)| = \begin{cases} n-1, & \text{if } \Gamma \text{ is not bipartite} \\ n-2, & \text{if } \Gamma \text{ is bipartite.} \end{cases}$$

Proof For each eigenvalue $\mu \in \sigma(\Delta_{\mathcal{G}})$ with $\mu \notin \{0, 2\}$ we obtain exactly one $\lambda_{\mu,k} \in \sigma_{\mathcal{V},k}(\Gamma)$. The assertion follows from the fact that both eigenvalues 0 and 2 (if present) are simple (Theorem 2.2). □

In summary, Theorem 5.4 and the deduced procedure tell us that the values of the vertex eigenfunctions at the vertices $\mathcal{V}$ of Γ are uniquely determined by the eigenvectors of the harmonic graph Laplacian matrix $\Delta_{\mathcal{G}}$ of the underlying combinatorial graph $\mathcal{G}$. In particular, this means that for a fixed eigenvalue $\mu \in \sigma(\Delta_{\mathcal{G}})$, all vertex eigenfunctions associated to $\lambda_{\mu,k}, k = 0, 1, \ldots$ assume the same values at the vertices. The behavior across the edges can be resolved by the transformed eigenvalues (5.6), delivering increasingly oscillating eigenfunctions for increasing k. These findings can be well illustrated by the following two, very simple, example graphs.

Example 5.8 Consider the diamond graph Γ_{dia} with equilateral edge length $\ell = 1$ as introduced in Sect. 2.4. The harmonic graph Laplacian of the underlying combinatorial graph $\mathcal{G}_{\text{dia}}$ is given by

$$\Delta_{\mathcal{G}_{\text{dia}}} = \begin{pmatrix} 1 & -\frac{1}{3} & -\frac{1}{3} & -\frac{1}{3} \\ -\frac{1}{2} & 1 & -\frac{1}{2} & 0 \\ -\frac{1}{3} & -\frac{1}{3} & 1 & -\frac{1}{3} \\ -\frac{1}{2} & 0 & -\frac{1}{2} & 1 \end{pmatrix}.$$

The eigenvalues of $\Delta_{\mathcal{G}_{\text{dia}}}$ are $\mu_1 = 0$, $\mu_2 = 1$, $\mu_3 = \frac{4}{3}$, $\mu_4 = \frac{5}{3}$, i.e., we have three eigenvalues $\mu_2, \mu_3, \mu_4 \notin \{0, 2\}$. Thus, for each k we obtain three vertex eigenvalues $\lambda_{\mu_2,k}, \lambda_{\mu_3,k}, \lambda_{\mu_4,k}$ that can be calculated by rule (5.6), see Fig. 5.1.

The corresponding vertex eigenfunctions are illustrated in Fig. 5.2. In accordance with (5.7), the oscillation of the eigenfunctions on each edge increases with increasing k. Since the restriction of $\phi_{\mu,k}$ to the vertices for a given μ is given by the eigenvectors Φ corresponding to μ, all eigenfunctions $\phi_{\mu,k}$ assume the same values at the vertices of the graph.

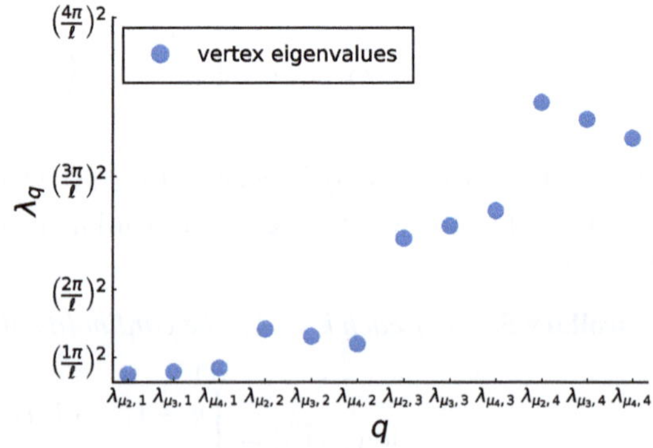

Fig. 5.1 Vertex eigenvalues of the diamond graph Γ_{dia} for $k = 0, 1, 2, 3$. *Note that the bunches come in increasing order although the ordering inside the bunches does not need to be increasing*

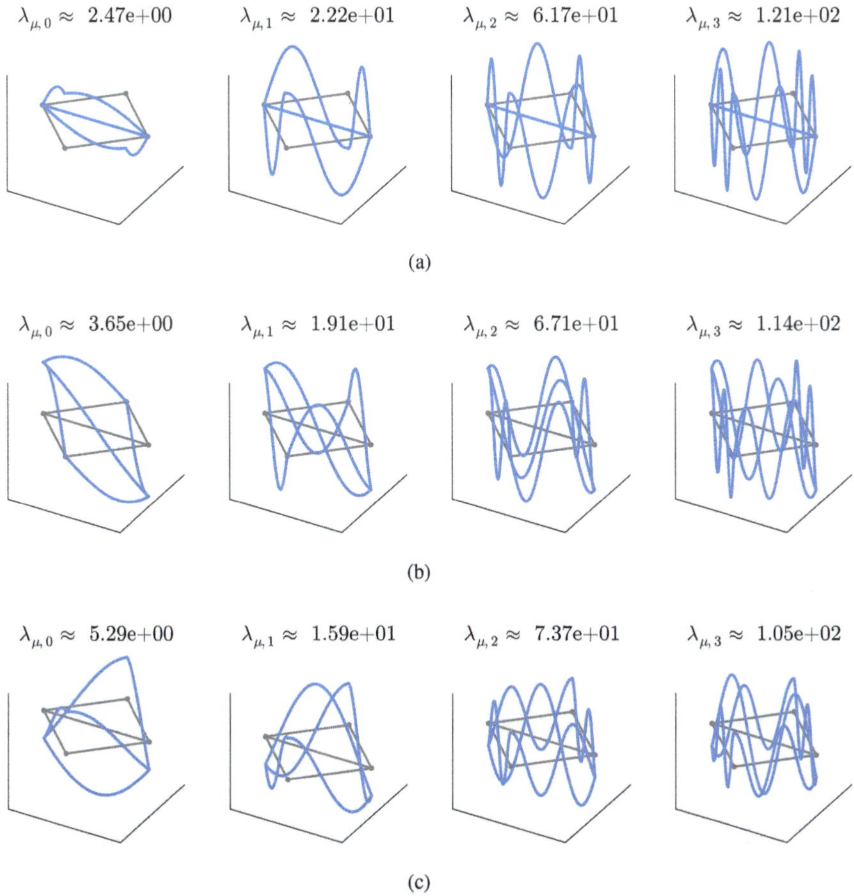

Fig. 5.2 Vertex eigenfunctions of Γ_{dia}. *Each row corresponds to a discrete eigenvalue μ and each column to $k = 0, 1, 2, 3$. Observe that the values of $\phi_{\mu,k}$ at the vertices of Γ_{dia} are the same for each k (for a fixed μ).* (**a**) Vertex Eigenfunctions $\phi_{\mu,k}$ for $\mu = 1$ and $k = 0, 1, 2, 3$. (**b**) Vertex Eigenfunctions $\phi_{\mu,k}$ for $\mu = \frac{4}{3}$ and $k = 0, 1, 2, 3$. (**c**) Vertex Eigenfunctions $\phi_{\mu,k}$ for $\mu = \frac{5}{3}$ and $k = 0, 1, 2, 3$

Example 5.9 Consider a star graph Γ_{star} with $n = 5$ vertices and equilateral edge length $\ell = 1$ as introduced in Sect. 2.4. The harmonic graph Laplacian

$$\Delta_{\mathcal{G}_{\text{star}}} = \begin{pmatrix} 1 & -\frac{1}{4} & -\frac{1}{4} & -\frac{1}{4} & -\frac{1}{4} \\ -1 & 1 & 0 & 0 & 0 \\ -1 & 0 & 1 & 0 & 0 \\ -1 & 0 & 0 & 1 & 0 \\ -1 & 0 & 0 & 0 & 1 \end{pmatrix}$$

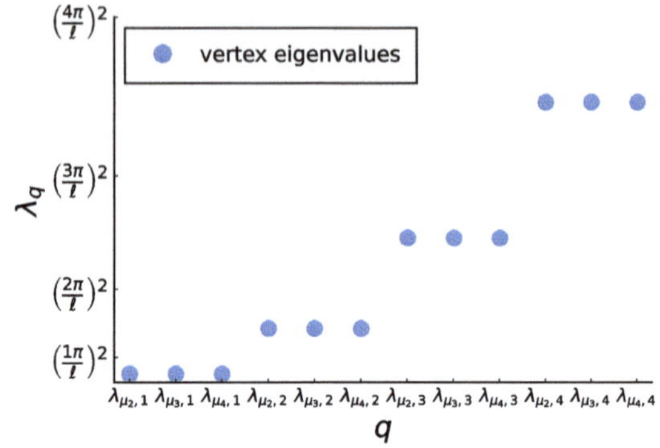

Fig. 5.3 Vertex eigenvalues of the star graph Γ_{star} for $k = 0, 1, 2, 3$

of the underlying combinatorial graph has eigenvalues $\mu_1 = 0, \mu_2 = 1, \mu_3 = 1, \mu_4 = 1, \mu_5 = 2$. Since Γ_{star} is bipartite, $2 \in \sigma(\Delta_{\mathcal{G}_{\text{star}}})$ and we have to exclude the two eigenvalues $\mu_1 = 0$ and $\mu_5 = 2$. Moreover, the only eigenvalue with $\mu \notin \{0, 2\}$ has multiplicity three. This is why every bunch $\sigma_{\mathcal{V},k}(\Gamma_{\text{star}})$ consists of three equal eigenvalues, see Fig. 5.3. Again, we observe that the eigenfunctions corresponding to the same discrete eigenvector assume the same values at the vertices, see Fig. 5.4.

5.2.2 *Non-vertex Spectrum*

We now turn to the so far neglected non-vertex spectrum, i.e., possible eigenvalues λ of Γ with $\sin\left(\sqrt{\lambda}\ell\right) = 0$. So far, we do not know if eigenvalues of this form exist. Clearly, if they exist, they have the form $\lambda = \left(\frac{k\pi}{\ell}\right)^2$ for $k \in \mathbb{N}_0$ and they need to fulfill (5.1). We start our investigation with the most trivial case $k = 0$.

Theorem 5.10 *Any given metric graph Γ has a simple eigenvalue $\lambda = 0$ corresponding to a constant eigenfunction.*

Proof For $\lambda = 0$, a general solution of (5.1) on each edge has the form

$$\phi_e(x) = A_e + B_e x.$$

The continuity condition implies

$$\phi_{\mathcal{V}}(v_i) = \phi_e(0) = A_e \quad \text{and} \quad \phi_{\mathcal{V}}(v_j) = \phi_e(\ell) = A_e + B_e\,\ell$$

for all $e = (v_i, v_j) \in \mathcal{E}$ where as usual $\phi_{\mathcal{V}}$ is the restriction of ϕ to the vertices. We can thus express ϕ_e in terms of the values at the vertices as

$$\phi_e(x) = \phi_{\mathcal{V}}(v_i) + \frac{\phi_{\mathcal{V}}(v_j) - \phi_{\mathcal{V}}(v_i)}{\ell} x.$$

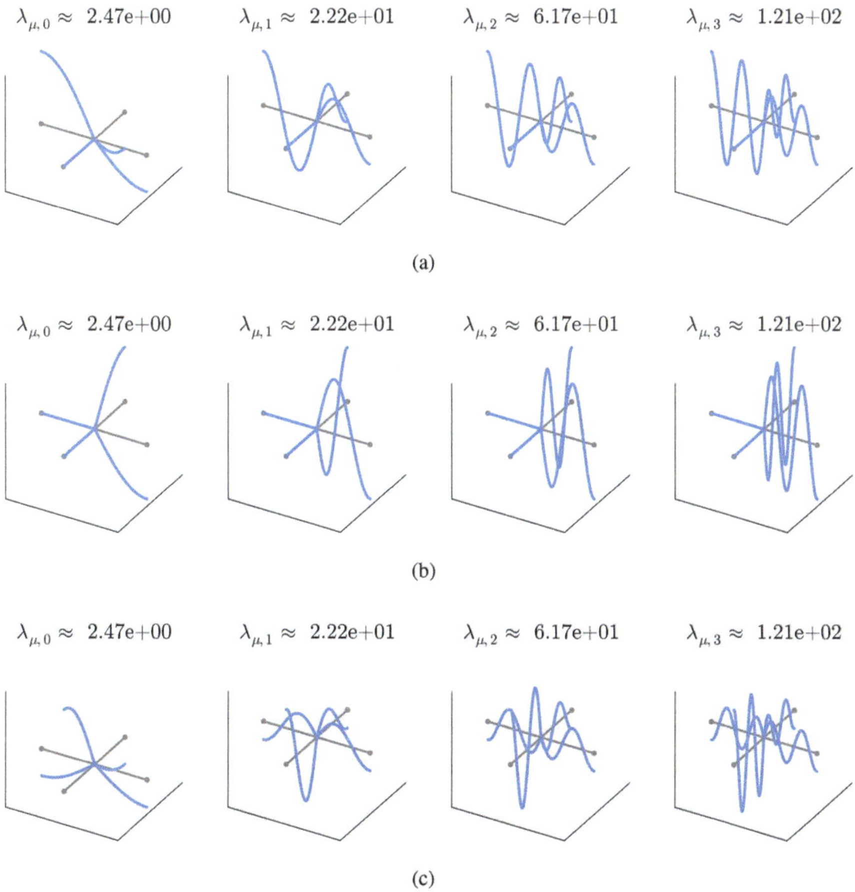

Fig. 5.4 Vertex eigenfunctions of Γ_{star}. *Note that although the eigenvalue μ is the same in each row, we observe three different eigenfunction patterns as a result of the three eigenvectors corresponding to μ.* (**a**) Vertex Eigenfunctions $\phi_{\mu,k}$ for $\mu = 1$ and $k = 0, 1, 2, 3$. (**b**) Vertex Eigenfunctions $\phi_{\mu,k}$ for $\mu = 1$ and $k = 0, 1, 2, 3$. (**c**) Vertex Eigenfunctions $\phi_{\mu,k}$ for $\mu = 1$ and $k = 0, 1, 2, 3$

Moreover, ϕ has to fulfill $\mathcal{K}(v_i) = 0$, i.e.,

$$\sum_{(v_i,v_j)\in\mathcal{E}} \frac{\phi_{\mathcal{V}}(v_j) - \phi_{\mathcal{V}}(v_i)}{\ell} = -\frac{\deg(v_i)}{\ell}\phi_{\mathcal{V}}(v_i) + \sum_{v_j\sim v_i} \frac{\phi_{\mathcal{V}}(v_j)}{\ell} = 0$$

for all $v_i \in \mathcal{V}$ which is equivalent to

$$\mathbf{L}\phi_{\mathcal{V}} = \mathbf{0}. \tag{5.8}$$

Here, $\mathbf{L}$ is the graph Laplacian matrix of the underlying combinatorial graph. According to Theorem 2.1, the only non-trivial solution of (5.8) is given by a constant $\phi_{\mathcal{V}}$, i.e., $\phi_{\mathcal{V}}(v_i) = \phi_{\mathcal{V}}(v_j)$ for all $v_i, v_j \in \mathcal{V}$. We conclude that

$$\phi_e(x) = \phi_{\mathcal{V}}(v_i) + \frac{\phi_{\mathcal{V}}(v_j) - \phi_{\mathcal{V}}(v_i)}{\ell} x = \phi_{\mathcal{V}}(v_i) + \frac{\phi_{\mathcal{V}}(v_i) - \phi_{\mathcal{V}}(v_i)}{\ell} x = \phi_{\mathcal{V}}(v_i)$$

for all $e \in \mathcal{E}$, i.e., ϕ is a constant function on Γ. □

For $\lambda > 0$, a general solution of (5.1) on each edge e has the form

$$\phi_e(x) = A_e \cos\left(\frac{k\pi}{\ell} x\right) + B_e \sin\left(\frac{k\pi}{\ell} x\right) \quad \text{for } x \in [0, \ell] \tag{5.9}$$

where the constants $A_e, B_e \in \mathbb{R}$ need to be determined. Using the continuity condition on the vertices, we observe the following first result on the behavior of possible eigenfunctions.

Lemma 5.11 *Let $k \in \mathbb{N}$, $\phi : \Gamma \to \mathbb{R}$ be defined by (5.9) for all $e \in \mathcal{E}$. Then, ϕ is continuous if and only if*

$$\phi_e(x) = \phi_{\mathcal{V}}(v_i) \cos\left(\frac{k\pi}{\ell} x\right) + B_e \sin\left(\frac{k\pi}{\ell} x\right)$$

where $\phi_{\mathcal{V}}$ is the restriction of ϕ to the vertices of Γ with

(a) if k is even: $\phi_{\mathcal{V}}$ is constant across the vertices,
(b) if k is odd: $\phi_{\mathcal{V}}(v_i) = -\phi_{\mathcal{V}}(v_j)$ for all adjacent vertices v_i, v_j. In particular, if Γ is not bipartite, the only possible choice is $\phi_{\mathcal{V}}(v) = 0$ for all $v \in \mathcal{V}$, i.e.,

$$\phi_e(x) = B_e \sin\left(\frac{k\pi}{\ell} x\right).$$

Proof

(a) For k even and an arbitrary edge $e = (v_i, v_j)$ it holds that

$$\phi_{\mathcal{V}}(v_i) = \phi_e(0) = A_e \cos(0) + B_e \sin(0) = A_e$$

and

$$\phi_{\mathcal{V}}(v_j) = \phi_e(\ell) = A_e \cos(k\pi) + B_e \sin(k\pi) = A_e.$$

Thus, $\phi_{\mathcal{V}}(v_i) = \phi_{\mathcal{V}}(v_j)$ for all $v_i, v_j \in \mathcal{V}$ with $v_i \sim v_j$. As Γ is connected, it follows that ϕ is constant on the vertices of Γ. Further, we derive

$$\phi_e(x) = \phi_{\mathcal{V}}(v_i) \cos\left(\frac{k\pi}{\ell}x\right) + B_e \sin\left(\frac{k\pi}{\ell}x\right).$$

(b) For k odd, the continuity condition reads $\phi_{\mathcal{V}}(v_i) = \phi_e(0) = A_e$ and $\phi_{\mathcal{V}}(v_j) = \phi_e(\ell) = -A_e$. This is fulfilled for $A_e = 0$ implying $\phi_{\mathcal{V}} = \mathbf{0}$ and $\phi_e(x) = B_e \sin\left(\frac{k\pi}{\ell}x\right)$. Moreover, if Γ is bipartite, we can choose $\phi_{\mathcal{V}}$ such that $\phi_{\mathcal{V}}(v_i) = -\phi_{\mathcal{V}}(v_j)$ for every two incident vertices v_i, v_j which leads to the second possible form of ϕ_e for bipartite graphs.

□

In particular, Lemma 5.11 tells us how the non-vertex eigenfunctions at the vertices of the graph must look like, namely

$$\phi_{\mathcal{V}}(v_i) = \phi_{\mathcal{V}}(v_j) \text{ for all } v_i, v_j \in \mathcal{V}$$

or

$$\phi_{\mathcal{V}}(v_i) = -\phi_{\mathcal{V}}(v_j) \text{ for all } v_i, v_j \in \mathcal{V}$$

where the last expression only has a nontrivial solution if Γ is bipartite. The remaining condition to satisfy is the conservation of currents condition $\mathcal{K}(v) = 0$ for all $v \in \mathcal{V}$. It turns out that there are several choices for the constants B_e on the edges delivering linear independent eigenfunctions ϕ fulfilling this requirement. One approach to study the number of possible choices of B_e is based on the idea of finding even and odd cycles in a graph and was presented in von Below's early work on eigenvalue problems on c^2-networks [53]. Here, however, we would like to take a less technical approach that works completely without the search for cycles in graphs and on top of that even delivers the values of B_e as well as the values of $\phi_{\mathcal{V}}$. To achieve this, we will make use of the extended graph, which was originally introduced for the discretization of Γ in Sect. 3.1.2, and the following simple observation given in [6].

Lemma 5.12 *Let v be a vertex with degree two and u be a function on Γ with $u_e \in H^2(e)$ for both edges e incident to v. Let further $\tilde{e}$ be the edge that emerges upon eliminating v and combining the two incident edges into one edge. Then, u satisfies the Neumann-Kirchhoff condition at v if and only if $u_{\tilde{e}} \in H^2(\tilde{e})$.*

Proof ([6], Remark 1.4.2) The continuity condition ensures the continuity of u, and the current conservation conditions the continuity of u' at v. Therefore, the two adjacent H^2-pieces of u match into one H^2-function on the combined edge. □

Consequently, the insertion of *artificial vertices* of degree two does not affect the solutions of the eigenvalue problem (5.1) on Γ. In this context, we now, with a slight abuse of notation, define the extended graph in the following sense.

Definition 5.13 Let Γ be a metric graph with edge length ℓ. Then, $\tilde{\Gamma}_k$ with edge length $\ell/(k+1)$ is the extended graph that arises from the insertion of k artificial vertices on each edge $e \in \mathcal{E}$.

Remark 5.14 We briefly state some properties of the extended graph we will occasionally need in the following.

(a) If Γ is equilateral, $\tilde{\Gamma}_k$ has equilateral edge length $\tilde{\ell} = \ell/(k+1)$.
(b) $\tilde{\Gamma}_k$ has $\tilde{n} = n + k \times m$ vertices.
(c) If the number of artificial vertices k on each edge is odd, $\tilde{\Gamma}_k$ is bipartite.

By Lemma 5.12, we have that the eigenfunctions of Γ agree with the eigenfunctions of the extended graph $\tilde{\Gamma}$. This turns out to be extremely useful due to the following simple yet important fact.

Lemma 5.15 *Every non-vertex eigenvalue* $\lambda = \left(\frac{k\pi}{\ell}\right)^2$ *of a metric graph* Γ *with length* ℓ *is a vertex eigenvalue of the extended graph* $\tilde{\Gamma}_k$ *with k artificial vertices on each edge.*

Proof $\tilde{\Gamma}_k$ has edge length $\tilde{\ell} = \frac{\ell}{k+1}$, i.e., $\sin\left(\sqrt{\lambda}\,\tilde{\ell}\right) = \sin\left(\frac{k}{k+1}\pi\right) \neq 0$. This means that λ is a vertex eigenvalue of $\tilde{\Gamma}_k$. □

If we denote by $\tilde{\Delta}_k := \Delta_{\tilde{\mathcal{G}}_k}$ the harmonic graph Laplacian of the underlying combinatorial graph, we can thus use the extended graph to again construct the non-vertex eigenfunctions by the eigenvectors of the discrete matrix $\tilde{\Delta}_k$.

Lemma 5.16 *Let* $k \in \mathbb{N}$, $\tilde{\Gamma}_k$ *be the extended graph to* Γ *with edge length* $\tilde{\ell}$ *and let* $\lambda = \left(\frac{k\pi}{\ell}\right)^2 \in \sigma_{\mathcal{E}}(\Gamma)$. *Then, the eigenfunctions corresponding to* λ *on the edges* $\tilde{e}$ *of the extended graph are given by*

$$\tilde{\phi}_{\tilde{e}}(x) = \frac{1}{\sin\left(\sqrt{\lambda}\tilde{\ell}\right)}\left(\tilde{\Phi}(v_i)\sin\left(\sqrt{\lambda}(\tilde{\ell}-x)\right) + \tilde{\Phi}(v_j)\sin\left(\sqrt{\lambda}x\right)\right)$$

where $\tilde{\Phi}$ *is an eigenvector of the harmonic graph Laplacian* $\tilde{\Delta}_k$ *corresponding to the eigenvalue* $\tilde{\mu} = 1 - \cos\left(\frac{k}{k+1}\pi\right)$.

Proof Follows directly from Lemma (5.15) and Theorem (5.4). □

On the original edges $e \in \mathcal{E}$, the eigenfunctions have the following form.

Theorem 5.17 *Let* $\lambda = \left(\frac{k\pi}{\ell}\right)^2$ *be a non-vertex eigenvalue of* Γ *and* $\tilde{\Phi}$ *be an eigenvector of the harmonic graph Laplacian* $\tilde{\Delta}_k$ *associated to* $\tilde{\mu} = \left(1 - \cos\left(\frac{k}{k+1}\pi\right)\right)$. *Then, a non-vertex eigenfunction of* Γ *corresponding to* λ *is given by*

$$\phi_e(x) = \tilde{\Phi}(v_i)\cos\left(\frac{k\pi}{\ell}x\right) + B_e \sin\left(\frac{k\pi}{\ell}x\right)$$

with

$$B_e = \frac{1}{\sin\left(\frac{k}{k+1}\pi\right)}\left(\tilde{\Phi}(v_{e,1}) - \tilde{\Phi}(v_i)\cos\left(\frac{k}{k+1}\pi\right)\right)$$

where $v_{e,1}$ is the first artificial vertex on edge e.

Proof We consider an arbitrary edge $e = (v_i, v_j)$ in the original graph Γ. In the extended graph $\tilde{\Gamma}_k$, this edge is partitioned into $k+1$ edges $(v_{e,0}, v_{e,1}), (v_{e,1}, v_{e,2}), \ldots, (v_{e,k}, v_{e,k+1})$ where $v_{e,0} := v_i$ and $v_{e,k+1} := v_j$. As λ is in the vertex-spectrum of $\tilde{\Gamma}_k$, we obtain by Lemma 5.16 that the corresponding eigenfunctions of the extended graph on each partition $(v_{e,r}, v_{e,r+1})$ have the form

$$\tilde{\phi}_{(v_{e,r},v_{e,r+1})} = \frac{1}{\sin\left(\sqrt{\lambda}\tilde{\ell}\right)}\left(\tilde{\Phi}(v_{e,r})\sin\left(\sqrt{\lambda}(\tilde{\ell}-x)\right) + \tilde{\Phi}(v_{e,r+1})\sin\left(\sqrt{\lambda}x\right)\right)$$

where $\tilde{\Phi}$ is an eigenvector of $\tilde{\Delta}_k$ corresponding to the eigenvalue $\tilde{\mu} = 1 - \cos\left(\frac{k}{k+1}\pi\right)$. On the other hand, by Lemma 5.12, we know that the eigenfunctions $\tilde{\phi}_{(v_{e,r},v_{e,r+1})}$ must agree with the non-vertex eigenfunctions ϕ_e of the original graph on each interval $\left[r\frac{\ell}{k+1}, (r+1)\frac{\ell}{k+1}\right]$, which, in turn, have to fulfill the continuity condition on the vertices. According to Lemma 5.11, a continuous function on edge e has the form

$$\phi_e(x) = \phi_{\mathcal{V}}(v_i)\cos\left(\frac{k\pi}{\ell}x\right) + B_e\sin\left(\frac{k\pi}{\ell}x\right)$$

where B_e is constant on the edge e. With the arguments above we obtain that ϕ is a non-vertex eigenfunction if

$$\phi_{\mathcal{V}}(v_i) = \phi_e\,(0) = \tilde{\phi}_{(v_i,v_{e,1})}\,(0) = \tilde{\Phi}(v_i)$$

$$\text{and} \qquad \phi_e\left(\frac{\ell}{k+1}\right) = \tilde{\phi}_{(v_i,v_{e,1})}(\tilde{\ell}) = \tilde{\Phi}(v_{e,1}),$$

i.e., $\tilde{\Phi}(v_{e,1}) = \phi_e\left(\frac{\ell}{k+1}\right) = \tilde{\Phi}(v_i)\cos\left(\frac{k}{k+1}\pi\right) + B_e\sin\left(\frac{k}{k+1}\pi\right)$ and consequently

$$B_e = \frac{1}{\sin\left(\frac{k}{k+1}\pi\right)}\left(\tilde{\Phi}(v_{e,1}) - \tilde{\Phi}(v_i)\cos\left(\frac{k}{k+1}\pi\right)\right).$$

□

As pointed out earlier in this section, the multiplicity of non-vertex eigenvalues of a metric graph was first treated in [53]. However, as already mentioned, a rather theoretical derivation using the occurrence of cycles in a graph is presented there. We will now demonstrate that our approach delivers a simple and straightforward proof for the multiplicity of non-vertex eigenvalues by only counting the eigenvalues of the extended harmonic graph Laplacian matrix. We start with the following lemma about the occurrence of vertex-eigenvalues in the extended graph.

Lemma 5.18 *The first k-bunches of vertex eigenvalues of* Γ *are contained in the first bunch of vertex eigenvalues of* $\tilde{\Gamma}$*, i.e.,*

$$\bigcup_{i=0}^{k} \sigma_{\mathcal{V},i}(\Gamma) \subset \sigma_{\mathcal{V},0}(\tilde{\Gamma}).$$

Proof We first point out again that Γ and $\tilde{\Gamma}$ have the same eigenvalues. By definition, $\sigma_{\mathcal{V},0}(\tilde{\Gamma})$ contains all vertex-eigenvalues of $\tilde{\Gamma}$ with $\lambda < \left(\frac{\pi}{\tilde{\ell}}\right)^2 = \left(\frac{\pi(k+1)}{\ell}\right)^2$. Hence, it has to contain all vertex-eigenvalues of Γ in $\bigcup_{i=0}^{k} \sigma_{\mathcal{V},i}(\Gamma)$. □

In other words, if $\Delta_{\mathcal{G}}$ has p eigenvalues μ with $\lambda \in \sigma_{\mathcal{V}}(\Gamma)$, the harmonic graph Laplacian of the extended graph with k artificial vertices has $(k+1)p$ such eigenvalues.

Corollary 5.19 *The multiplicity of* $\lambda = \left(\frac{k\pi}{\ell}\right)^2$ *as an eigenvalue of* Γ *is given by*

$$\begin{cases} 1, & \text{if } k = 0 \\ m - n + 2, & \text{if } k \text{ even} \\ m - n, & \text{if } k \text{ odd and } \Gamma \text{ not bipartite} \\ m - n + 2, & \text{if } k \text{ odd and } \Gamma \text{ bipartite} \end{cases}$$

where n is the number of vertices and m the number of edges of Γ.

Proof We first suppose that Γ is not bipartite. For $k = 0$, it follows from Theorem 5.10 that a constant function across all edges is the only eigenfunction corresponding to $\lambda = 0$. Hence, the multiplicity of 0 as eigenvalue is one. In total, $\tilde{\Delta}_k = \Delta_{\mathcal{G}}$ has n eigenvalues. Since Γ is not bipartite, $2 \notin \sigma(\Delta_{\mathcal{G}})$. We deduce that the remaining $n-1$ eigenvalues of $\Delta_{\mathcal{G}}$ correspond to the vertex spectrum of Γ. The multiplicities for $k > 0$ follow with the following observations for $k = 1$ and $k = 2$:

$(k = 1)$: The extended graph has $n+m$ vertices, which means that $\tilde{\Delta}_k$ has $n+m$ eigenvalues. We know that 0 is an eigenvalue of $\tilde{\Delta}_k$ with multiplicity one and that $\tilde{\Delta}_k$ has $2\,(n-1)$ eigenvalues belonging to $\sigma_{\mathcal{V}}(\Gamma)$. Moreover, as k is odd, $\tilde{\Gamma}_k$ is bipartite and therefore 2 appears in the spectrum of $\tilde{\Delta}_k$ corresponding to the quantum graph eigenvalue $\lambda = \left(\frac{2\pi}{\ell}\right)^2$. There remain $(n+m)-(1+1+2n-2) = m-n$ eigenvalues belonging to $\lambda = \left(\frac{\pi}{\ell}\right)^2$.

$(k = 1)$: The extended graph has $n + 2m$ vertices. As elaborated for the case $k = 1$, $\tilde{\Delta}_k$ has eigenvalue $\mu = 0$ with multiplicity 1 and $3(n-1)$ vertex eigenvalues. Moreover, $\tilde{\Delta}_k$ has $(m-n)$ eigenvalues corresponding to $\left(\frac{\pi}{\ell}\right)^2$ and we already found $\lambda = \left(\frac{2\pi}{\ell}\right)^2$ with multiplicity 1 in $\tilde{\Delta}_{k-1}$. The remaining $(n+2m) - (1 + 1 + 3n - 3 + m - n) = (m-n) + 1$ eigenvalues also belong to $\lambda = \left(\frac{2\pi}{\ell}\right)^2$ which gives us a total of $m - n + 2$.

If Γ is bipartite, $\mu = 2$ is eigenvalue of $\Delta_{\mathcal{G}}$. Therefore, $\tilde{\Delta}_k$ has $(k+1)(n-2)$ eigenvalues belonging to the vertex spectrum of Γ and the claim follows with the same arguments. □

Thus, our results agree with the investigations of von Below in [53] although they originate from a rather practical approach. It is now time to revisit the two example graphs from the previous subsection and illustrate their non-vertex spectrum.

Example 5.20 Consider Γ_{dia} with equilateral edge length $\ell = 1$ and its extended graph $\tilde{\Gamma}_1$ with equilateral edge length $\tilde{\ell} = \ell/2 = 1/2$. The extended graph has $\tilde{n} = n + m = 9$ vertices and its harmonic graph Laplacian matrix has nine eigenvalues

$$\tilde{\mu}_1 = 0,\ \tilde{\mu}_2 = \frac{-\sqrt{2}+2}{2},\ \tilde{\mu}_3 = \frac{-\sqrt{3}+3}{3},\ \tilde{\mu}_4 = \frac{-\sqrt{6}+6}{6},$$
$$\tilde{\mu}_5 = 1,\ \tilde{\mu}_6 = \frac{\sqrt{6}+6}{6},\ \tilde{\mu}_7 = \frac{\sqrt{3}+3}{3},\ \tilde{\mu}_8 = \frac{\sqrt{2}+2}{2},\ \tilde{\mu}_9 = 2.$$

Clearly, we have one simple zero eigenvalue and $\tilde{\mu} = 2$ since $\tilde{\Gamma}_1$ is bipartite. Applying formula (5.6) for $k = 0$ and $\tilde{\ell} = \ell/2$ to the eigenvalues $\tilde{\mu}_2, \tilde{\mu}_3, \tilde{\mu}_4$ delivers exactly the three vertex eigenvalues in the first bunch of $\sigma_{\mathcal{V},0}(\Gamma_{\text{dia}})$ of the original graph. It follows one non-vertex eigenvalue $\lambda = \left(\frac{\pi}{\ell}\right)^2 = \left(\frac{1}{\tilde{\ell}}\arccos(1 - \tilde{\mu}_5)\right)^2$ and the second bunch of vertex eigenvalues of the original graph corresponding to $\tilde{\mu}_6, \tilde{\mu}_7, \tilde{\mu}_8$.

The same considerations apply for growing extended graphs which we exemplary illustrate in Fig. 5.5 for $\tilde{\Gamma}_k, k = 1, 2, 3$. The results are in accordance with the expected multiplicity of non-vertex eigenvalues we proved in Corollary 5.19, namely, for k odd, we obtain $m - n = 1$ and for k even $m - n + 2 = 3$ non-vertex eigenvalues.

Moreover, the eigenvectors of the extended harmonic graph Laplacian $\tilde{\Delta}_1$ deliver the values of the non-vertex eigenfunctions on the vertices $\mathcal{V}_1$ of $\tilde{\Gamma}_1$ which can be interpolated over the edges using rule (5.7), see Fig. 5.6. As in the vertex case, we observe that the oscillation of ϕ across the edges increases for growing k. Moreover, for k odd, the eigenfunction is zero on the vertices of the original graph and for k eve $\phi_{\mathcal{V}}$ is constant.

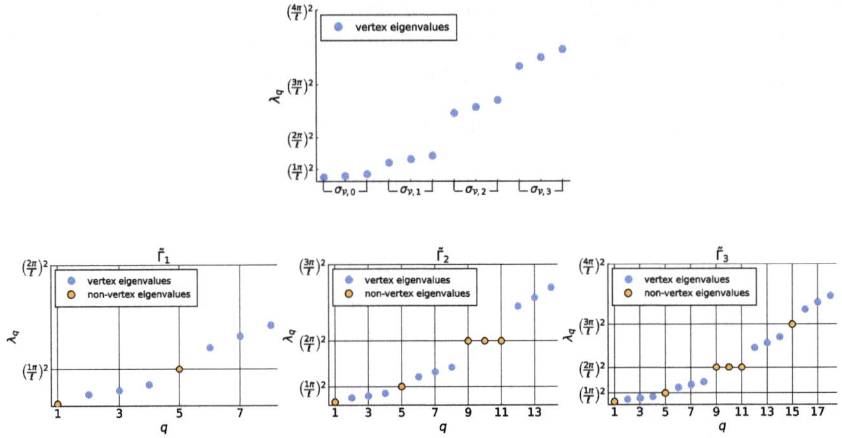

Fig. 5.5 Vertex eigenvalues of the extended graphs of Γ_{dia}. *To facilitate comparison, we plotted the first four bunches of vertex eigenvalues of* Γ_{dia} *again in the first row. In the second row, we see the first bunch of vertex eigenvalues for the extended graphs* $\tilde{\Gamma}_1, \tilde{\Gamma}_2, \tilde{\Gamma}_3$. *We emphasized the non-vertex eigenvalues of* Γ_{dia} *in a different color. Observe that* $\sigma_{\mathcal{V},0}(\tilde{\Gamma}_3)$ *contains all eigenvalues in* $\bigcup_{i=0}^{3} \sigma_{\mathcal{V},i}(\Gamma_{\text{dia}})$ *plus five non-vertex eigenvalues* $\lambda_5 = \pi^2, \lambda_9 = \lambda_{10} = \lambda_{11} = (2\pi)^2$ *and* $\lambda_{15} = (3\pi)^2$

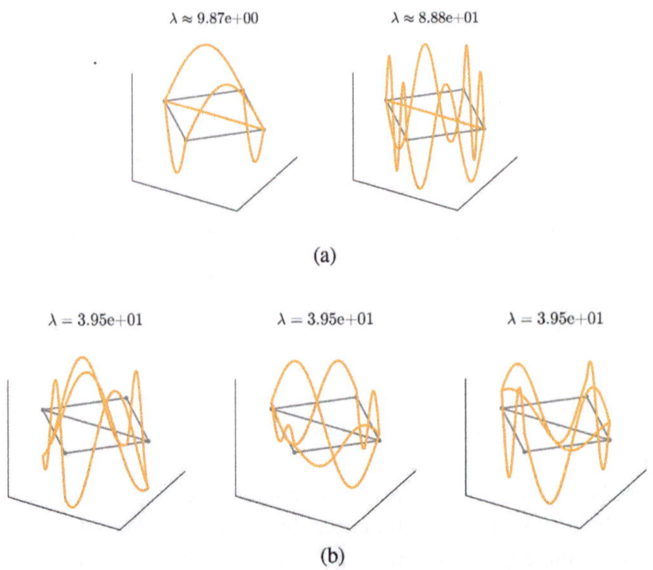

Fig. 5.6 Non-vertex eigenfunctions of Γ_{dia} separated by k odd and even. *Note that the constant eigenfunction corresponding to the first non-vertex eigenvalue* $\lambda = 0$ *is not shown. The second non-vertex eigenfunction is given for* $k = 1$, *i.e.,* $\lambda = \pi^2$ *in the left image of the upper row. This function is zero on the vertices of the original graph, just like the second illustrated non-vertex eigenfunction for k odd in the right image of the upper row. For k even, the non-vertex eigenvalue* $(2\pi)^2$ *occurs with multiplicity three and has three linear independent associated eigenfunctions that are given in the lower part of the figure*. **(a)** k odd, $k = 1, 3$. **(b)** k even, $k = 2$

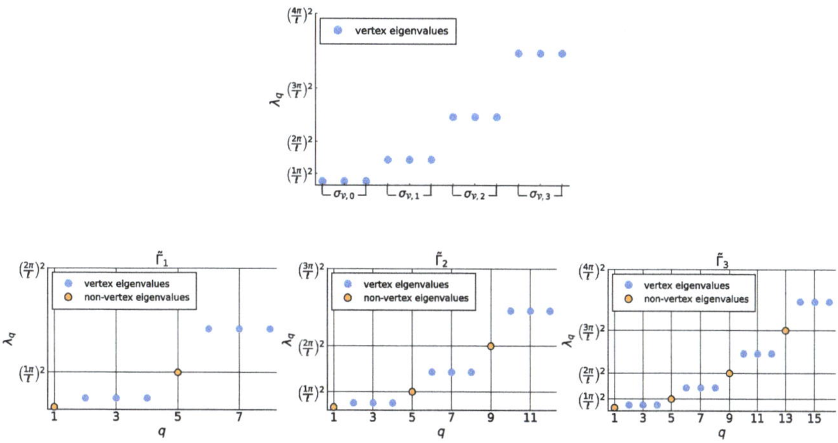

Fig. 5.7 Vertex eigenvalues of the extended graph of Γ_{star}. *To facilitate comparison, we plotted the first four bunches of vertex eigenvalues of* Γ_{dia} *again in the first row. In the second row, we see the first bunch of vertex eigenvalues for the extended graphs* $\tilde{\Gamma}_1, \tilde{\Gamma}_2, \tilde{\Gamma}_3$. *We emphasized the non-vertex eigenvalues of* Γ_{dia} *in a different color. Observe that* $\sigma_{V,0}(\tilde{\Gamma}_3)$ *contains all eigenvalues in* $\bigcup_{i=0}^{3} \sigma_{V,i}(\Gamma_{\text{dia}})$ *plus three non-vertex eigenvalues* $\lambda_5 = \pi^2$, $\lambda_9 = (2\pi)^2$ *and* $\lambda_{13} = (3\pi)^2$

Example 5.21 We repeat the previous discussion for the bipartite star graph Γ_{star} with equilateral edge length $\ell = 1$. The extended graph $\tilde{\Gamma}_1$ of Γ_{star} has $\tilde{n} = n+m = 9$ vertices and $\tilde{m} = 2m = 8$ edges with equilateral edge length $\tilde{\ell} = \ell/2 = 1/2$. The harmonic graph Laplacian of $\tilde{\Gamma}_1$ has eigenvalues

$$\tilde{\mu}_1 = 0, \tilde{\mu}_2 = \tilde{\mu}_3 = \tilde{\mu}_4 = \frac{-\sqrt{2}+2}{2}, \tilde{\mu}_5 = 1, \tilde{\mu}_6 = \tilde{\mu}_7 = \tilde{\mu}_8 = \frac{\sqrt{2}+2}{2}, \tilde{\mu}_9 = 2.$$

Since $k = 1$ is odd, $\tilde{\Gamma}_1$ is bipartite and we obtain the two simple eigenvalues $\tilde{\mu}_1 = 0$ and $\tilde{\mu}_9 = 2$. Other than this, we have $n - m + 2 = 1$ odd non-vertex eigenvalues arising from $\tilde{\mu}_5$ and the first two bunches of vertex eigenvalues of the original graph corresponding to $\tilde{\mu}_2, \tilde{\mu}_3, \tilde{\mu}_4$ and $\tilde{\mu}_6, \tilde{\mu}_7, \tilde{\mu}_8$. In Fig. 5.7, we illustrate this behavior for $\tilde{\Gamma}_1, \tilde{\Gamma}_2$ and $\tilde{\Gamma}_3$. Again, $\bigcup_{i=0}^{3} \sigma_{V,i}(\Gamma_{\text{star}}) \subseteq \sigma_{V,0}(\tilde{\Gamma}_3)$, and we observe the multiplicity of the non-vertex eigenvalues to be one, which also follows from Corollary 5.19 and the fact that Γ_{star} is bipartite.

In Fig. 5.8, we plotted the non-vertex eigenfunctions for $k = 1, 2, 3$. We are now in the interesting case that Γ_{star} is bipartite and can thus observe the alternating behavior of the eigenfunctions for k odd ($\lambda \approx 9.87\text{e}+00$ and $\lambda \approx 8.88\text{e}+01$). For k even, the same holds true as for the non-bipartite case, i.e., the eigenfunction assumes the same value at all the vertices.

To conclude this section, we want to emphasize that according to our definition and notation, the eigenvalues come in increasing order with growing k. In particular, the non-vertex eigenvalues exactly match in between the bunches of vertex eigenvalues.

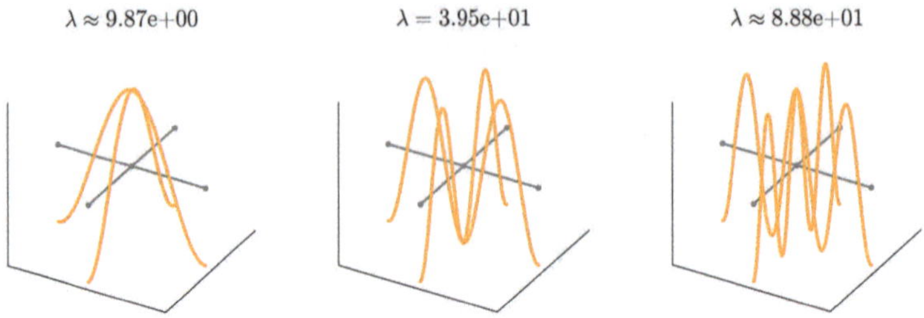

Fig. 5.8 Non-vertex eigenfunctions of Γ_{star} for $k = 1, 2, 3$. *Observe that for k odd, the eigenfunction assumes alternating values at the vertices*

5.2.3 Aspects of Implementation

We first summarize the above considerations to compute an arbitrarily large lower part of the spectrum with corresponding eigenfunctions in Algorithm 4 and, for the rest of this subsection, discuss the implementation in detail step by step. The algorithm is constructed in such a way that we obtain an ascending sequence of eigenvalues with associated eigenfunctions.

We start with $k = 0$ and the simple zero eigenvalue $\lambda = 0$ with corresponding constant eigenfunction in line 4. Next, in line 12–15, the first bunch of vertex eigenfunctions for $k = 0$ is computed and delivers all eigenvalues $\lambda \in \left(0, \left(\frac{\pi}{\ell}\right)^2\right)$. We recall that here and for all the following k, the number of computed vertex eigenfunctions is given by

$$\#\phi^{\mathcal{V}} = \begin{cases} n-1, & \text{if } \Gamma \text{ is not bipartite} \\ n-2, & \text{if } \Gamma \text{ is bipartite.} \end{cases} \tag{5.10}$$

We now raise k to one and compute the non-vertex eigenfunctions corresponding to $\lambda = \left(\frac{\pi}{\ell}\right)^2$ in line 18–21. In Lemma 5.15, we observed that they appear as vertex eigenfunctions in the extended graph $\tilde{\Gamma}_k$ and thus can be constructed using the eigenvectors $\tilde{\Phi}$ of $\tilde{\Delta}_k$ corresponding to the eigenvalue $\tilde{\mu} = \left(1 - \cos\left(\frac{k}{k+1}\pi\right)\right)$, compare Lemma 5.16. These eigenvectors are given as the nontrivial solutions of

$$\tilde{\Delta}_k \tilde{\Phi} = \tilde{\mu} \tilde{\Phi}.$$

For k odd, the number of non-trivial solutions, i.e., the number of eigenfunctions is given by

$$\#\phi^{\mathcal{E},\text{odd}} = \begin{cases} m-n, & \text{if } \Gamma \text{ is not bipartite} \\ m-n+2, & \text{if } \Gamma \text{ is bipartite,} \end{cases}$$

Algorithm 4 Compute all eigenfunctions of Γ corresponding to $\lambda < \left(\frac{k_{\max}\pi}{\ell}\right)^2$

1: Compute all eigenpairs (μ, Φ) of $\Delta_{\mathcal{G}}$
2: **for** $k = 0, \ldots, k_{\max} - 1$ **do**
3: **if** k even **then**
4: **if** $k = 0$ **then** ▷ non-vertex eigenfunction (ϕ_0)
5: $\lambda = 0, \phi \equiv 1$
6: **else** ▷ non-vertex eigenfunctions ($\phi^{\mathcal{E},\text{even}}$)
7: **for** $\tilde{\Phi}$ with $\tilde{\Delta}_k \tilde{\Phi} = \left(1 - \cos\left(\frac{k}{k+1}\pi\right)\right)\tilde{\Phi}$ **do**
8: $\lambda = \left(\frac{k\pi}{\ell}\right)^2$
9: $\phi_e = \tilde{\Phi}(v_i)\cos\left(\sqrt{\lambda}\,x\right) + \frac{1}{\sin\left(\frac{k}{k+1}\pi\right)}\left(\tilde{\Phi}(v_{e,1}) - \tilde{\Phi}(v_i)\cos\left(\frac{k}{k+1}\pi\right)\right)\sin\left(\sqrt{\lambda}\,x\right)$
10: **end for**
11: **end if**
12: **for** (μ, Φ) with $\mu \in \sigma(\Delta_{\mathcal{G}})\backslash\{0, 2\}$ **do** ▷ vertex eigenfunctions ($\phi^{\mathcal{V}}$)
13: $\lambda = \left(\frac{1}{\ell}\left(\arccos(1-\mu) + k\pi\right)\right)^2$,
14: $\phi_e = \frac{1}{\sin(\sqrt{\lambda}\,\ell)}(\Phi(v_i)\sin(\sqrt{\lambda}(\ell - x)) + \Phi(v_j)\sin(\sqrt{\lambda}x))$.
15: **end for**
16: **end if**
17: **if** k odd **then** ▷ non-vertex eigenfunctions ($\phi^{\mathcal{E},\text{odd}}$)
18: **for** $\tilde{\Phi}$ with $\tilde{\Delta}_k \tilde{\Phi} = \left(1 - \cos\left(\frac{k}{k+1}\pi\right)\right)\tilde{\Phi}$ **do**
19: $\lambda = \left(\frac{k\pi}{\ell}\right)^2$
20: $\phi_e = \tilde{\Phi}(v_i)\cos\left(\sqrt{\lambda}\,x\right) + \frac{1}{\sin\left(\frac{k}{k+1}\pi\right)}\left(\tilde{\Phi}(v_{e,1}) - \tilde{\Phi}(v_i)\cos\left(\frac{k}{k+1}\pi\right)\right)\sin\left(\sqrt{\lambda}\,x\right)$
21: **end for**
22: **for** (μ, Φ) with $\mu \in \sigma(\Delta_{\mathcal{G}})\backslash\{0, 2\}$ **do** ▷ vertex eigenfunctions ($\phi^{\mathcal{V}}$)
23: $\lambda = \left(\frac{1}{\ell}\left(-\arccos(1-\mu) + (k+1)\pi\right)\right)^2$,
24: $\phi_e = \frac{1}{\sin(\sqrt{\lambda}\,\ell)}(\Phi(v_i)\sin(\sqrt{\lambda}(\ell - x)) + \Phi(v_j)\sin(\sqrt{\lambda}x))$.
25: **end for**
26: **end if**
27: **end for**

see Corollary 5.19. Subsequently, the second bunch of vertex eigenfunctions corresponding to $\lambda \in \left(\left(\frac{\pi}{\ell}\right)^2, \left(\frac{2\pi}{\ell}\right)^2\right)$, where the number of eigenfunctions is again given by (5.10), is computed in line 22–25.

For $k = 2$, we enter the for loop in line 6–10 for the first time and compute the non-vertex eigenfunctions corresponding to $\lambda = \left(\frac{2\pi}{\ell}\right)^2$ using the extended graph equivalent to the considerations for k odd. The number of obtained eigenfunctions is given by

$$\#\phi^{\mathcal{E},\text{even}} = m - n + 2$$

according to Corollary 5.19.

Eigenfunctions—Storage and Normalization

In practice, it is often not necessary and also not recommended to assemble the single eigenfunctions as indicated in Algorithm 4. Instead, if we express both the vertex and non-vertex eigenfunctions in the form $\phi_e = A_e \cos\left(\sqrt{\lambda}x\right) + B_e \sin\left(\sqrt{\lambda}x\right)$, we only have to compute and store the constants A_e, B_e for each edge. The non-vertex eigenfunctions are already provided in this form and we recall that we have

$$A_e^{\mathcal{E},k} = \tilde{\Phi}(v_i), \quad B_e^{\mathcal{E},k} = \frac{1}{\sin\left(\frac{k}{k+1}\pi\right)}\left(\tilde{\Phi}(v_{e,1}) - \tilde{\Phi}(v_i)\cos\left(\frac{k}{k+1}\pi\right)\right).$$

For the vertex eigenvalues, we note that

$$\begin{aligned}\phi_e^{\mathcal{V}}(x) &= \frac{1}{\sin\left(\sqrt{\lambda}\,\ell\right)}\left(\Phi(v_i)\sin\left(\sqrt{\lambda}(\ell - x)\right) + \Phi(v_j)\sin\left(\sqrt{\lambda}x\right)\right)\\ &= \Phi(v_i)\cos(\sqrt{\lambda}x) + \frac{1}{\sin(\sqrt{\lambda}\,\ell)}\left(\Phi(v_j) - \Phi(v_i)\cos(\sqrt{\lambda}\,\ell)\right)\sin(\sqrt{\lambda}x)\end{aligned}$$

(as seen in the proof of Theorem 5.2) and thus obtain

$$A_e^{\mathcal{V},\lambda} = \Phi(v_i), \quad B_e^{\mathcal{V},\lambda} = \frac{1}{\sin(\sqrt{\lambda}\,\ell)}\left(\Phi(v_j) - \Phi(v_i)\cos(\sqrt{\lambda}\,\ell)\right).$$

With this considerations in place, it is also trivial to compute the norms of the eigenfunctions, namely as

$$(\phi, \phi)_\Gamma = \sum_{e\in\mathcal{E}} \int_e \left(A_e \cos\left(\sqrt{\lambda}x\right) + B_e \sin\left(\sqrt{\lambda}x\right)\right)^2 \mathrm{d}x$$

where

$$\begin{aligned}&\int_e \left(A_e \cos\left(\sqrt{\lambda}x\right) + B_e \sin\left(\sqrt{\lambda}x\right)\right)^2 \mathrm{d}x\\ &\quad= \frac{2(A_e^2 + B_e^2)\sqrt{\lambda}\,\ell + (A_e^2 - B_e^2)\sin(2\sqrt{\lambda}\,\ell) + 2A_eB_e(1 - \cos(2\sqrt{\lambda}\ell))}{4\sqrt{\lambda}}.\end{aligned}$$

Eigendecomposition of the Harmonic Graph Laplacian

A computationally expensive part of Algorithm 4 is the eigendecomposition of the harmonic graph Laplacian $\Delta_{\mathcal{G}}$, which, in particular, is not symmetric. However, we pointed out in Sect. 2.1.2 that $\Delta_{\mathcal{G}}$ is similar to the symmetric normalized graph Laplacian $\mathcal{L}$. Thus, whenever we need to compute the eigenpairs (μ, Φ) of $\Delta_{\mathcal{G}}$, we proceed as follows:

1. Compute normalized graph Laplacian matrix $\mathcal{L} = \mathbf{D}^{-\frac{1}{2}}\mathbf{L}\mathbf{D}^{-\frac{1}{2}}$
2. Compute all eigenpairs (μ, Υ) of $\mathcal{L}$
3. Compute $\Phi = \mathbf{D}^{-\frac{1}{2}}\Upsilon$

Eigenvectors of the Extended Graph Laplacian Matrices

Of special interest are further the eigenvectors of the extended harmonic graph Laplacian $\tilde{\Delta}_k$ corresponding to the eigenvalue $\tilde{\mu} = \left(1 - \cos\left(\frac{k\pi}{\ell}\,\tilde{\ell}\right)\right) = \left(1 - \cos\left(\frac{k}{k+1}\pi\right)\right)$. Clearly, we do not need to compute a complete eigendecomposition of $\tilde{\Delta}_k$ but only need to find $\tilde{\Phi}$ with $\tilde{\Delta}_k\tilde{\Phi} = \tilde{\mu}\,\tilde{\Phi}$ using, e.g., a Krylov-subspace method.

5.3 Spectra of Non-equilateral Graphs

The following subsection is based on modified and unmodified passages of the manuscript [18]. The idea and the manuscript were developed in joint work with Chong-Son Dröge, a master student working under my supervision. We both contributed equally to the conceptualization and the formulation of the manuscript as well as the design and implementation of numerical experiments. However, for this work, I have partly changed the outline and structure in order to give more examples and a more detailed derivation of the ideas that led to the proposed ansatz.

As derived in Sect. 5.1, the computation of quantum graph vertex eigenvalues for non-equilateral metric graphs can be reduced to solving a *Nonlinear Eigenvalue Problem* (NEP): Find $\lambda > 0$ such that there exists a nontrivial $\Phi \in \mathbb{R}^n$ with

$$\mathbf{H}(\lambda)\Phi = \mathbf{0} \tag{5.11}$$

where $\mathbf{H}(z) \in \mathbb{R}^{n\times n}$ is defined by

$$\mathbf{H}_{ij}(z) := \begin{cases} -\sum_{e\in\mathcal{E}_{v_i}} \cot\left(\sqrt{z}\,\ell_e\right), & \text{if } i = j \\ \frac{1}{\sin(\sqrt{z}\,\ell_e)}, & \text{if } e = (v_i, v_j) \in \mathcal{E} \\ 0 & \text{otherwise.} \end{cases}$$

Clearly, (5.11) has a nontrivial solution Φ only if $\mathbf{H}(z)$ is singular. In other words, we are looking for $z > 0$ with

$$\det(\mathbf{H}(z)) = 0 \tag{5.12}$$

and we are in search of an efficient method to compute the roots of $\det(\mathbf{H}(z))$. Note that we introduced z to differentiate between the variable z and the eigenvalue λ.

In the context of NEPs, various methods have been developed to tackle this problem, see for example [26] for a detailed review. In particular, a common ansatz are methods based on linearization which include an approximation of (5.11) by a polynomial NEP.

We want to present a different strategy here by exploiting the fact that the spectrum of equilateral graphs can be computed in a highly efficient way. The approach relies on the assumption that we can find a *good* approximation of a non-equilateral graph by an equilateral graph. The eigenvalues of this *equilateral approximation* are then used as initial guesses for the numerical solution of (5.12).

The computation of non-equilateral quantum graph eigenvalues using an approximation with an equilateral graph has also been proposed in [32], yet not in the context of the NEP (5.11), see also Sect. 5.3.4. I wish to point out that our approach was developed independently and the manuscript [32] was only found by us afterwards.

To support the assumption that non-equilateral graphs can be approximated by equilateral ones, the following exposition will be preceded by some motivational thoughts and examples. These are intended to help the reader following the subsequent presentation. We then briefly outline a Newton-trace iteration to find the roots of (5.12). In the main part of this subsection, we discuss different types of equilateral approximations and how to utilize them to generate initial guesses for the Newton-trace iteration.

We point out that the following survey is to be understood as a first attempt to solve our problem. We are convinced that the development of more sophisticated methods leading to more efficient solvers for our type of NEPs is an exciting objective of future work.

5.3.1 Motivation

As we have already learned in Lemma 5.12, the elimination or insertion of vertices of degree two on the edges does not influence the spectrum of the quantum graph. Intuitively, the idea arises to approximate Γ by an extended graph with equilateral edge length, as illustrated in Fig. 5.9. The spectrum of this equilateral graph can then easily be computed by the method proposed in the previous section.

Clearly, if Γ has integer edge lengths as in the illustration, we cannot only find an approximation but an exact representation of Γ by constructing an extended graph with equilateral edge length according to the greatest common divisor of the edge

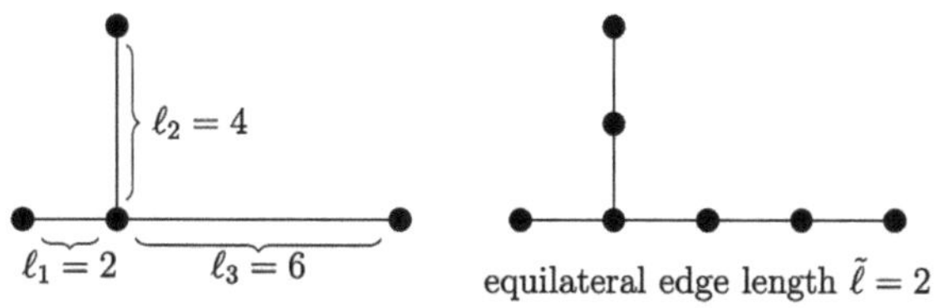

Fig. 5.9 Extended equilateral graph [18]

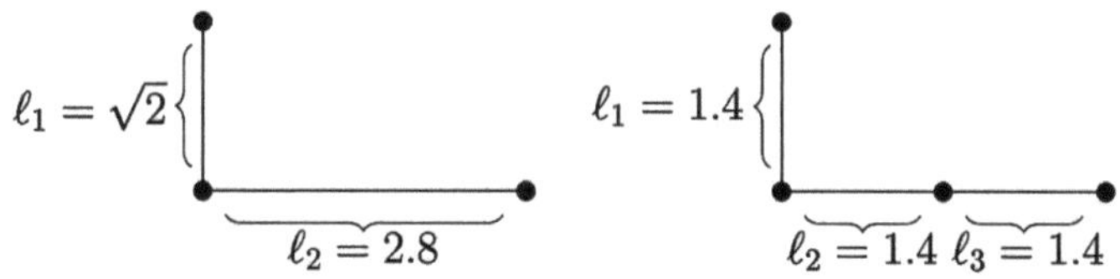

Fig. 5.10 Equilateral approximation graph [18]

lengths. In the worst case, this is one resulting in a large extended graph, especially when the range of the edge lengths is large. A similar strategy can be applied to any graph with rational edge lengths. However, if the greatest common divisor strategy results in a too large expansion as well as for graphs with general edge lengths $\ell_e \in \mathbb{R}$, we have to work with an approximation instead of an exact representation Γ, see for example Fig. 5.10.

Even if it appears to be reasonable at first, we cannot assume that these approximations also provide good estimates of the eigenvalues. Changing the edge lengths could cause new effects in the spectrum and deviations from the patterns found for equilateral graphs. We therefore start this section with a simple motivational experiment: Given an initial graph Γ with equilateral edge length, we construct a graph Γ' with the same underlying combinatorial graph but with slightly different and, in particular, non-equilateral edge lengths. We then compare a fixed part of their lower eigenvalues to investigate the effect of the deviations on the spectrum.

Let us start with the simple graph Γ_{dia} with $n = 4$ vertices and $m = 5$ edges and assign a basic edge length $\ell = 1$ to all edges. As a first modified graph Γ'_{dia}, we choose a copy of Γ_{dia} and only vary the length of edge $\ell_2 = 1$ to $\ell'_2 = 1.1$ as illustrated in Fig. 5.11. Note that the spectrum of Γ'_{dia} can be calculated exactly by constructing its extended graph with equilateral edge length 0.1. In the upper left plot of Fig. 5.12, we compare the first 12 eigenvalues of Γ_{dia} and Γ'_{dia}. The spectra of three other variations Γ'_{dia} are illustrated in the remaining figures, where the edge lengths of Γ'_{dia} are modified as indicated in the legends.

Before we turn to the discussion of the findings in Fig. 5.12, we repeat the experiment with a bipartite graph, the star graph Γ_{star} with $n = 5$ and $m = 4$, see Fig. 5.13.

In general, for all four scenarios of lengths variation, the smaller eigenvalues of Γ can be approximated by Γ' much better than the larger eigenvalues. If only one edge length is modified (i.e., upper row of Figs. 5.13 and 5.12), the difference between

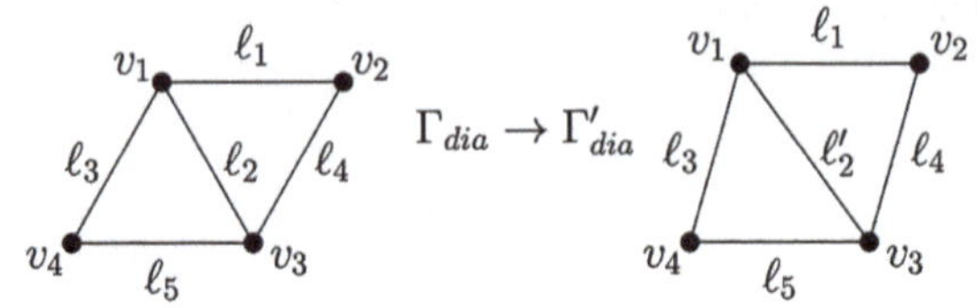

Fig. 5.11 Diamond graph with varied edge length. *The edge length of* ℓ_2 *is varied in* Γ'_{dia}

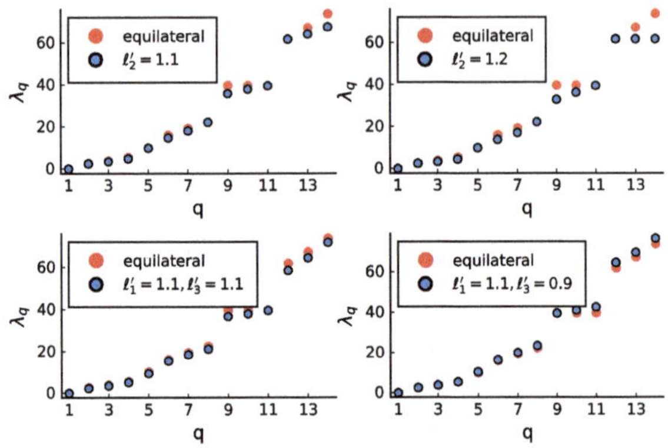

Fig. 5.12 Patterns in equilateral and non-equilateral graph spectra of Γ_{dia}. *Each plot contains the eigenvalues for variations of the edge lengths as indicated in the legend. Note that the red dots are not visible beyond the blue ones if the eigenvalues agree*

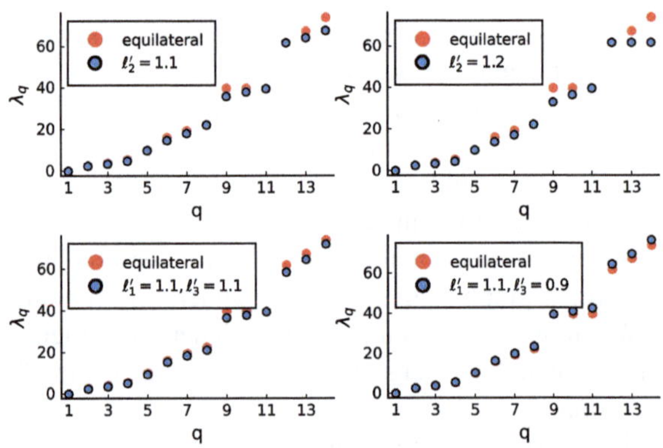

Fig. 5.13 Patterns in equilateral and non-equilateral graph spectra of Γ_{star}. *Each plot contains the eigenvalues for variations of the edge lengths as indicated in the legend. Note that the red dots are not visible beyond the blue ones if the eigenvalues agree*

the eigenvalues of Γ and Γ' grows with increasing distance of the edge lengths. Interestingly, the stretching of the edge in both examples causes the eigenvalues to become smaller. This is consistent with what we expect from a diffusion process on a graph with enlarged edge lengths: the state of equilibrium is reached later.

If we vary more than one edge length, not only the values of the eigenvalues show a higher deviation but also the patterns are further disturbed. If we enlarge one edge and simultaneously shrink another one, the patterns seem to match "better" again. However, we can now observe a deviation in both directions, i.e., some eigenvalues are larger and some are smaller. In the experiments before, where we exclusively enlarged the edge length, we observed that the eigenvalues of Γ always approximate the eigenvalues of Γ' from below. In fact, this observation will later be a key assumption and instrument for our proposed algorithm.

5.3.2 Newton-Trace Iteration

Let us now turn back to the nonlinear eigenvalue problem (5.11) which can be reformulated as the root-finding problem (5.12). Since we have to evaluate $\mathbf{H}$ for different z frequently, we first consider an efficient way to assemble $\mathbf{H}(z)$.

Lemma 5.22 *For a given metric graph Γ with edge lengths $\boldsymbol{\ell}$ and incidence matrix $\mathbf{N}$, the matrix $\mathbf{H}$ corresponding to the NEP (5.11) is given by*

$$\mathbf{H}(z) = \operatorname{diag}\left(\left\{\left(\mathbf{N}\mathbf{W}_1(z)\mathbf{N}^T\right)_{ii}\right\}_{i=1}^{n}\right) - \mathbf{N}\mathbf{W}_2(z)\mathbf{N}^T + \operatorname{diag}\left(\left\{\left(\mathbf{N}\mathbf{W}_2(z)\mathbf{N}^T\right)_{ii}\right\}_{i=1}^{n}\right)$$

with

$$\mathbf{W}_1(z) = \operatorname{diag}\left(\left\{-\cot\left(\sqrt{z}\,\ell_e\right)\right\}_{e\in\mathcal{E}}\right) \text{ and } \mathbf{W}_2(z) = \operatorname{diag}\left(\left\{-1/\sin\left(\sqrt{z}\,\ell_e\right)\right\}_{e\in\mathcal{E}}\right).$$

Proof We separate $\mathbf{H}(z)$ in $\mathbf{H}_1(z) + \mathbf{H}_2(z)$ where $\mathbf{H}_1(z)$ contains the diagonal and $\mathbf{H}_2(z)$ the off-diagonal entries. $\mathbf{H}_2(z)$ is the weighted adjacency matrix of Γ with edge weights $w_2(e) = \frac{1}{\sin(\sqrt{z}\,\ell_e)}$. It can be described as $\mathbf{H}_2(z) = -\mathbf{N}\mathbf{W}_2(z)\mathbf{N}^T + \operatorname{diag}\left(\left\{\left(\mathbf{N}\mathbf{W}_2(z)\mathbf{N}^T\right)_{ii}\right\}_{i=1}^{n}\right)$ with $\mathbf{W}_2(z)$ as given above. On the other hand, the matrix $\mathbf{H}_1(z)$ containing the diagonal entries is the degree matrix of the graph Γ with edge weights $w_1(e) = -\cot\left(\sqrt{z}\,\ell_e\right)$ and thus given as $\mathbf{H}_1(z) = \operatorname{diag}\left(\left\{\left(\mathbf{N}\mathbf{W}_1(z)\mathbf{N}^T\right)_{ii}\right\}_{i=1}^{n}\right)$ with $\mathbf{W}_1(z)$ defined in the assertion. □

However, it would be very optimistic to tackle this problem with a classical iterative root-finding algorithm since we are facing a highly nonlinear problem, compare Fig. 5.14. And, of course, each iteration of a classical iterative solver is very expensive due to the evaluation of the determinant.

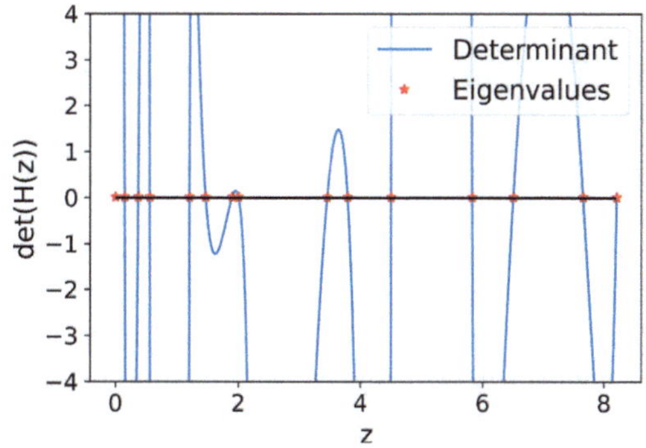

Fig. 5.14 Determinant of $\mathbf{H}(z)$ for $z \in [0, 9]$ for a random example graph with 5 vertices and 6 edges. *The course of the determinant is plotted together with the true eigenvalues indicated with an asterix. Observe that the roots of the determinant agree with the eigenvalues (illustration from [17])*

To overcome the latter, [26] describes one popular approach to compute the roots of (5.12) by the Newton-trace method given in [39]. This can be easily derived by considering the standard Newton iteration for $\det(\mathbf{H}(z))$ given by

$$z^{j+1} = z^j - \frac{\det(\mathbf{H}(z^j))}{(\det(\mathbf{H}(z^j)))'}.$$

For $\det(\mathbf{H}(z)) \neq 0$, the trace theorem (e.g. [38], p. 390, (7)) allows us to find the derivative of $\det(\mathbf{H}(z))$ as

$$\det(\mathbf{H}(z))' = \det(\mathbf{H}(z))\,\mathrm{trace}(\mathbf{H}^{-1}(z)\,\mathbf{H}'(z))$$

provided that the entries of $\mathbf{H}(z)$ are differentiable functions of z. The Newton iteration then simplifies to the Newton-trace iteration

$$z^{j+1} = z^j - \frac{1}{\mathrm{trace}(\mathbf{H}^{-1}(z^j)\,\mathbf{H}'(z^j))}. \tag{5.13}$$

The key to an efficient application of (5.13) are suitable initial guesses. This is in particular important for our problem since we are searching for more than one root and we do not know the amount of the roots in a given interval a priori. As indicated at the beginning of this section, our objective will be to apply the eigenvalues of equilateral approximations to start the Newton iteration.

Remark 5.23 In fact, (5.13) is still a rather expensive iteration and there exist several other approaches to apply modified Newton methods for our problem, for example based on QR-decomposition or a nonlinear inverse iteration applied directly to the vector equation (5.11), see again [26] for a review. However, we will restrict ourselves to the simple Newton-trace iteration and in the remainder of this section focus on the derivation of initial guesses to ensure fast convergence.

5.3.3 Approximation via Equilateral Graphs

In this section, we concretize the ideas presented in the motivation in Sect. 5.3.1 and discuss some numerical findings on the computation of non-equilateral graph spectra by equilateral approximations and the subsequent improvement via Newton-trace iterations. The main result of this subsection is the formulation of a *Nested Iteration Algorithm* to efficiently compute initial guesses. In practice, the main challenge will be to find the best trade-off between the accuracy of the initial guesses and the number of required Newton-trace iterations.

The following definition adapted from [37] is useful for our investigations.

Definition 5.24 For any metric graph Γ, we define its *cleaned graph* $\text{clean}(\Gamma)$ by removing all vertices of degree two and combining the two incident edges to one combined edge.

The cleaned graph $\text{clean}(\Gamma)$ and Γ have the same spectrum but not the same underlying combinatorial graph $\mathcal{G}$. Let us define an equilateral approximation of Γ as follows.

Definition 5.25 Let Γ be a non-equilateral metric graph with edge lengths $\boldsymbol{\ell}$ and underlying combinatorial graph $\mathcal{G}$. An *equilateral approximation* of Γ is a graph $\mathfrak{G}_h$ with equilateral edge length h such that $\text{clean}(\mathfrak{G}_h)$ has underlying combinatorial graph $\mathcal{G}$ and edge lengths $\mathfrak{l}_h \approx \boldsymbol{\ell}$.

We further introduce the distance of Γ and its equilateral approximation $\mathfrak{G}_h$.

Definition 5.26 Let Γ be a non-equilateral, clean metric graph with edge lengths $\boldsymbol{\ell}$ and $\mathfrak{G}_h$ an equilateral approximation of Γ. Then, we define the *distance* between Γ and $\mathfrak{G}_h$ as

$$\text{dist}(\Gamma, \mathfrak{G}_h) = \|\boldsymbol{\ell} - \mathfrak{l}_h\|$$

with $\|\cdot\| = \|\cdot\|_2$.

Note that Γ and $\text{clean}(\mathfrak{G}_h)$ have the same edges, i.e., $\boldsymbol{\ell} - \mathfrak{l}_h$ is well defined.

The main question is now whether the eigenvalues of $\mathfrak{G}_h$ converge to the eigenvalues of Γ for $\text{dist}(\Gamma, \mathfrak{G}_h) \to 0$. In the scope of this work, we only tackle this question from a numerical perspective and thus are in need of a sequence of equilateral approximations $\{\mathfrak{G}_h\}_{h\in\mathbb{R}^+}$ such that $\text{dist}(\Gamma, \mathfrak{G}_h) \to 0$ for $h \to 0$ for a given metric graph Γ.

As a first attempt, we propose the following heuristic:

---

for $J = 1, 2, \dots$ **do**
  Set $h = 2^{-J}$
  Initialize $\mathfrak{l} \in \mathbb{R}^m$
  for $e \in \mathcal{E}$ **do**
    Compute $N_e = \mathtt{round}(\ell_e/h)$
    Set $(\mathfrak{l})_e = h \times N_e$
  end for
  Compute $\mathfrak{G}_h$ as extended graph of $(\mathcal{V}, \mathcal{E})$ with equilateral edge length h
end for

In other words, our aim is to approximate each edge e by subdivisions of length h and we choose the number of subdivisions N_e for each edge such that $h \times N_e \geq \ell_e$ or $h \times N_e \leq \ell_e$, depending on which choice delivers a cleaned edge length closer to the original length ℓ_e, compare Fig. 5.15. In this context, we also speak of a *rounded equilateral approximation* and denote the resulting graphs by $\mathfrak{G}_h$ according to the lengths of the subdivisions, i.e., the lengths of the edges in $\mathfrak{G}_h$.

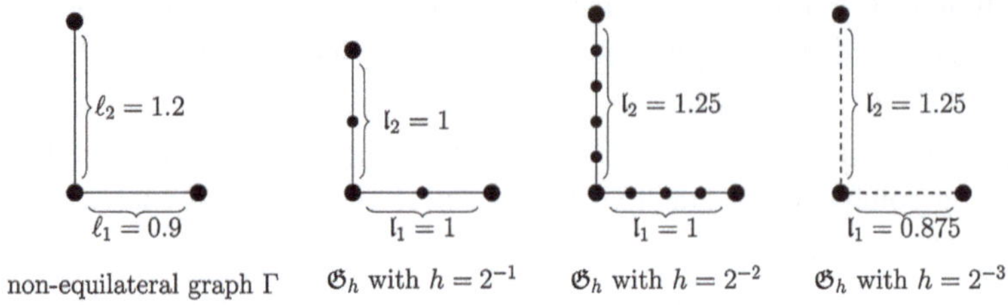

Fig. 5.15 Rounded equilateral approximations

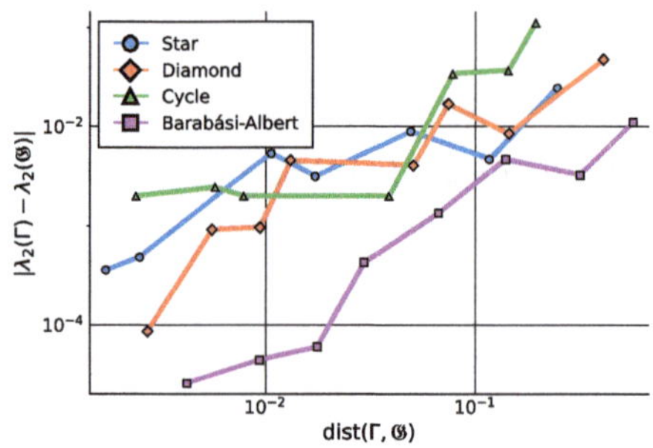

Fig. 5.16 Approximating the spectral graph of the four example graphs from Example 5.27 via rounded equilateral approximations. *The distance between the graphs is plotted together with the absolute error of the eigenvalue estimate*

Let us start with a first numerical experiment comparing the spectra of Γ and $\mathfrak{G}_h$ constructed according to the heuristic above for $h \to 0$.

Example 5.27 We consider four example graphs Γ_{dia}, Γ_{star}, Γ_{circle} and Γ_{BA} where Γ_{circle} is a cycle graph with $n = 4$ vertices and $m = 4$ edges and Γ_{BA} was generated with $n = 10$ vertices and $d = 3$. We equip the edges of each of the graphs with a random length between 1 and 2, rounded to three decimal digits. The exact eigenvalues of the non-equilateral graphs can thus be calculated by subdividing each edge in pieces of length 0.001. In Fig. 5.16, we illustrate the absolute difference between the first non-zero eigenvalue $\lambda_2(\Gamma)$ of the exact graph and the first non-zero eigenvalue $\lambda_2(\mathfrak{G}_h)$ of the equilateral approximations $\mathfrak{G}_h$ for $h = 2^{-J}$, $J = 1, \ldots, 9$.

We thereby already found an example showing that $|\lambda_2(\Gamma) - \lambda_2(\mathfrak{G}_h)| \nleq |\lambda_2(\Gamma) - \lambda_2(\mathfrak{G}_{h'})|$ for $h < h'$, i.e., that eigenvalues of closer graphs are not necessary better approximations. In our second approach, we want to learn from the observations in the motivational examples where we have seen that the eigenvalues seem to be approximated from below or above depending on whether the edges of the approximating graph are longer or shorter than the original ones. This motivates the following definition.

Definition 5.28 An equilateral approximation of Γ is referred to as *equilateral floor approximation* if $\mathfrak{l}_h - \ell \leq \mathbf{0}$, i.e., if any edge length of the approximation is shorter than the edge length of the exact graph. Equivalently, we use the terminology *equilateral ceil approximation* if all edge lengths are longer.

The computation of floor and ceil equilateral approximations can be easily done by a slight modification in the previous heuristic and is summarized in Algorithms 5 and 6.

Algorithm 5 Floor approximation

```
for J = 1, 2, ... do
    Set h = 2^{-J}, Initialize l ∈ R^m
    for e ∈ E do
        Compute N_e = floor(ℓ_e/h)
        Set (l)_e = h × N_e
    end for
    Compute extended graph G_{fl,h}
end for
```

Algorithm 6 Ceil approximation

```
for J = 1, 2, ... do
    Set h = 2^{-J}, Initialize l ∈ R^m
    for e ∈ E do
        Compute N_e = ceil(ℓ_e/h)
        Set (l)_e = h × N_e
    end for
    Compute extended graph G_{ce,h}
end for
```

And, in fact, repeating the previous example with a sequence of floor and ceil approximations shows better approximation properties.

Example 5.29 Consider the four example graphs Γ_{dia}, Γ_{star}, Γ_{circle} and Γ_{BA} from Example 5.27 with randomly chosen edge lengths between 1 and 2, rounded to three decimal digits. The approximation quality of the first non-zero eigenvalue $\lambda_2(\Gamma)$ by an equilateral floor and ceil approximation are displayed in Fig. 5.17.

Moreover, we compared the first 50 eigenvalues of the exact graphs with the first 50 eigenvalues of $\mathfrak{G}_{\text{fl},h}$ and $\mathfrak{G}_{\text{ce},h}$ and exemplary illustrate the absolute error $|\lambda_i(\Gamma) - \lambda_i(\mathfrak{G}_h)|$ for $h = 2^{-4}, 2^{-6}, 2^{-8}$ and all example graphs in Fig. 5.18. For each graph, the ceil approximations approach the eigenvalues from below while the eigenvalues of floor approximation from above. Consistent with Fig. 5.17, the approximation quality increases for decreasing h and thereby decreasing distance $\text{dist}(\Gamma, \mathfrak{G}_h)$. Higher eigenvalues exhibit a larger deviation as we have already seen in the motivational examples.

From the computational side, there arise several challenges we will address in the remainder of this section.

We need to keep in mind that although the equilateral eigenvalue problem can be solved exactly, its complexity massively grows for $h \to 0$. The previous section has

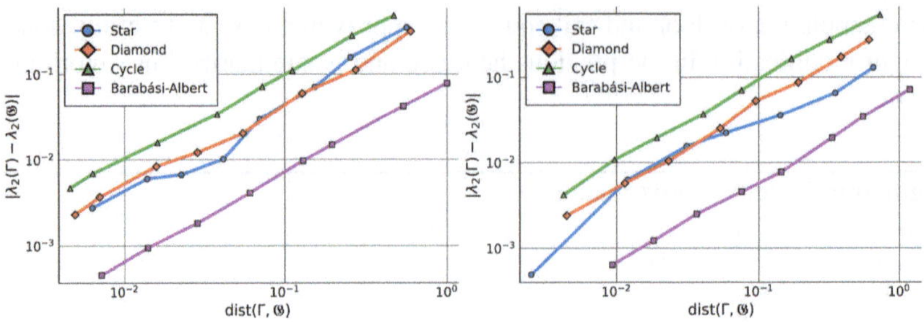

Fig. 5.17 Approximating the spectral gap of the four example graphs from Example 5.29 via equilateral floor and ceil approximations. *The distance between the graphs is plotted together with the absolute error of the eigenvalue estimate*

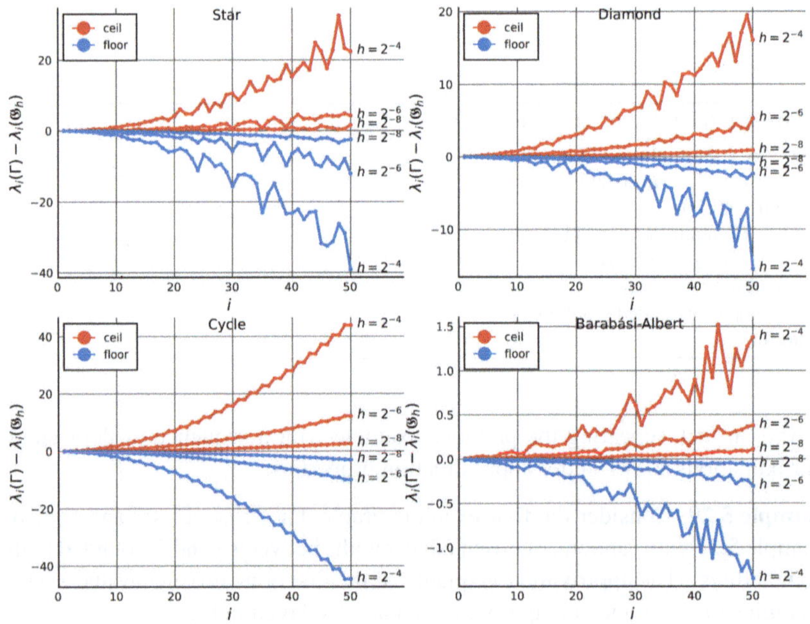

Fig. 5.18 Eigenvalues estimates of equilateral floor and ceil approximations of the four example graphs 5.29. *The absolute error* $|\lambda_i(\Gamma) - \lambda_i(\mathfrak{G}_h)|$ *for the first 50 eigenvalues is illustrated for* $h = 2^{-4}, 2^{-6}, 2^{-8}$ *using equilateral ceil and floor approximations. Observe that the ceil approximations approach the eigenvalues from below and the floor approximation from above*

shown that the equilateral eigenvalues are given by the (transformed) eigenvalues of the normalized graph Laplacian matrix of $\mathfrak{G}_h$. If we denote by $\mathfrak{L}_h$ the normalized graph Laplacian matrix of $\mathfrak{G}_h$, this means we have to solve the discrete eigenvalue problem

$$\mathfrak{L}_h \Phi_h = \mu_h \Phi_h.$$

Approximating the edges Γ with a lot of subdivisions N_e massively enlarges this system, quickly leading to the limits of a classical eigenvalue solver. First numerical experiments even showed that, for small h, standard `julia` or `python` implementations of sparse solvers fail to converge within a moderate number of iterations already for simple graphs like the star graph from the previous example.

Our objective is to overcome this problem with a *nested iteration* approach. This method is originally known from the solution of systems of linear equations arising from the discretization of PDEs. In brief, the idea is to use the solution obtained by a discretization with a coarser grid as starting vector for an iterative solver applied to the larger objective system. It seems to be natural to apply a similar strategy here since our hypothesis is that the eigenvalues of $\mathfrak{G}_h$ are a good approximation of the eigenvalues of $\mathfrak{G}_{h/2}$, especially for $h \to 0$.

To formalize the idea, recall that since $\mathfrak{G}_h$ is equilateral, the following relation between the eigenvalues of $\mathfrak{G}_h$ and $\mathfrak{L}$ derived in Theorem 5.3 holds true:

$$\mu_h = 1 - \cos\left(\sqrt{\lambda(\mathfrak{G}_h)}\, h\right).$$

Supposing $\lambda(\mathfrak{G}_h)$ is a good approximation for $\lambda(\mathfrak{G}_{h/2})$, then $1 - \cos\left(\sqrt{\lambda(\mathfrak{G}_h)}\, h/2\right)$ should be a good approximation of $\mu_{h/2}$. Therefore, we want to apply it as shift value σ for an inverse iteration to find the eigenvalues of $\mathfrak{L}_{h/2}$.

The nested iteration eigenvalue algorithm is summarized in Algorithm 7. Note that in the practical implementation, we always searched the three eigenvalues closest to the shift value in order to guarantee, that no eigenvalues are missed.

Recall that the actual goal was to utilize equilateral approximations to generate suitable initial guesses for a Newton-trace iteration and not as stand-alone approximation. It remains to investigate how good the approximations via equilateral graphs need to be to obtain a fast converging Newton-trace iteration. We address this question with two numerical experiments in Sect. 6.2.3.

5.3.4 *Alternative Approaches and Outlook*

Although not in the context of NEPs, the approximation of non-equilateral graph spectra by equilateral graphs has been studied in [32] and an error estimate has been derived under the assumption that the total length $\mathrm{vol}(\Gamma)$ of the non-equilateral graph is preserved by its equilateral approximation. However, the construction of a total length preserving approximation that can be represented by an equilateral extended graph leads to a constrained optimization problem, a so-called Simultaneous Diophantine Approximation (SDAP). In contrast, our approach relies on a fast and easy heuristic to compute a sequence of approximation graphs, which, combined with a nested iteration approach, can be utilized to develop an efficient method to approximate non-equilateral graph eigenvalues by equilateral graphs. Moreover, the subsequent Newton-iteration guarantees the convergence to

Algorithm 7 Nested iteration eigenvalue algorithm

Input metric graph Γ, number of eigenvalues Q, startlevel J_0
Output $\lambda_{\mathrm{fl},q}$ and $\lambda_{\mathrm{cl},q}$, $q = 1, \ldots, Q$

For $h = 2^{-J_0}$, compute first Q eigenvalues of $\mathfrak{G}_{\mathrm{fl},h}$ and $\mathfrak{G}_{\mathrm{cl},h}$ by sparse eigenvalue solver
for $J = J_0 + 1, J_0 + 2, \ldots$ **do**
  Set $h = 2^{-J}$
  Compute floor and ceil graphs $\mathfrak{G}_{\mathrm{fl},h}$ and $\mathfrak{G}_{\mathrm{cl},h}$ by Algorithm 5, 6
  for $q = 1, \ldots, Q$ **do**
    Compute eigenvalues $\mu_{\mathrm{fl},q}$ of $\mathfrak{L}_{\mathrm{fl},h}$ and $\mu_{\mathrm{cl},q}$ of $\mathfrak{L}_{\mathrm{cl},h}$ by inverse shift iteration with

$$\sigma_{\mathrm{fl}} = \left(1 - \cos\left(\sqrt{\lambda_{\mathrm{fl},q}^{(J-1)}}\right)\right)$$

    and

$$\sigma_{\mathrm{cl}} = \left(1 - \cos\left(\sqrt{\lambda_{\mathrm{cl},q}^{(J-1)}}\right)\right)$$

    Compute quantum graph eigenvalues

$$\lambda_{\mathrm{fl},q}^{J} = \left(\frac{1}{h}\left(\arccos(1 - \mu_{\mathrm{fl},q})\right)\right)^2 \text{ and } \lambda_{\mathrm{cl},q}^{J} = \left(\frac{1}{h}\left(\arccos(1 - \mu_{\mathrm{cl},q})\right)\right)^2$$

  end for
end for

an eigenvalue and relieves the necessity of constructing arbitrary good equilateral approximations. Nevertheless, it will be subject to future research to derive similar error bounds for our type of floor and ceil approximations and to compare the different approximation techniques in practice.

On the other hand, we can also choose a classical approach to solve the NEP (5.11) without the equilateral graph approximations at all. This could be a method based on linearization where **H** is, for instance, interpolated by Chebyshev polynomials. The resulting polynomial NEP can be reduced to a generalized linear eigenvalue problem and the solutions of this problem should in turn be good approximations of the original NEP and therefore can be applied to start, for example, the Newton-trace iteration.

At this point, we want to address the algorithm proposed by Brio et al. in [10], that we already outlined in the introduction, again. In principle, the idea also explores that the solution of the eigenvalue problem on each edge has the form $\phi_e(x) = A_e \cos(\sqrt{\lambda}\, x) + B_e \sin(\sqrt{\lambda}\, x)$ (as in our deviation of the NEP). However, they choose an edge based approach and impose two conditions per edge (corresponding to the Neumann-Kirchhoff conditions) to determine the coefficients A_e and B_e. In fact, this also results in a nonlinear eigenvalue problem but with a coefficient matrix $M(z)$ of size $2m \times 2m$ (which, since usually $m \gg n$, is computationally more expensive than ours). The algorithm they propose to solve this problem, besides a rather laborious assembling of the coefficient matrix, involves at one step to plot the reciprocal condition number of $M(z)$ and to graphically estimate the lower bound of the spacing between the eigenvalues. Thereupon, this is applied to define subintervals on which a line minimization algorithm can find λ as the roots of the reciprocal condition number. However, due to the manual processing required, we did not further compare it numerically to our proposed method.

Last but not least, yet another totally different ansatz is the vertex scattering approach. In a nutshell, a $2m \times 2m$ bond scattering matrix $S\left(\sqrt{\lambda}\right)$ describes the scattering of waves at the vertices of the graph [6]. The eigenvalues λ can then be found as the solutions of the *secular equation* $\det\left(\mathbf{I} - S\left(\sqrt{\lambda}\right)\mathrm{e}^{\mathrm{i}\sqrt{\lambda}\mathrm{L}}\right) = 0$. In this expression, L is a diagonal matrix containing the edge lengths (twice, since each edge is considered as bond with an in- and outgoing direction). In [48], it is claimed that though being of size $2m \times 2m$, the structure of the coefficient matrix in the scattering approach admits some favorable properties and, in fact, a spectral counting function to compute the number of eigenvalues less than or equal to a given value K is derived. However, the spectral counting function in each call requires a diagonalization of the bond scattering matrix at wave number K. It will be interesting to investigate if this approach can be exploited in the context of our algorithm.

Chapter 6
Numerical Results

Abstract The theoretical investigations of the previous chapters are verified, underlined and extended by various numerical experiments. This chapter includes numerical experiments on the convergence of the finite element semidiscretization and the Crank-Nicolson multigrid method. Further discussions on the computation of equilateral and non-equilateral quantum graph spectra are included as well as numerical results in the context of spectral solutions such as on the decay of coefficients and simulations of heat, fractional diffusion and reaction-diffusion equations. Finally, we compare the finite element and spectral Galerkin methods in terms of their achieved accuracy in relation to their complexity.

6.1 Finite Element Discretization

We start with some simple convergence experiments for elliptic and parabolic test problems and discuss the CN-MGM solution of the finite element semidiscretization.

6.1.1 Convergence of Finite Element Semidiscretization

As introductory examples, we consider a finite element discretization of Test Problem 2.20 and Test Problem 2.21. Those problems are of the form

$$\mathcal{H}u + u = f \tag{6.1}$$

subject to Neumann-Kirchhoff conditions on the star graph Γ_{star} and the diamond graph Γ_{dia}. The finite element discretization of (6.1) with step size h leads to a system of linear equations

$$\hat{\mathbf{L}}_h \mathrm{u}_h + \hat{\mathbf{M}}_h \mathrm{u}_h = \hat{\mathbf{f}}_h \tag{6.2}$$

A. Weller, *Numerical Methods for Metric Graphs*, Lecture Notes in Mathematics 2382, https://doi.org/10.1007/978-3-032-05011-3_6

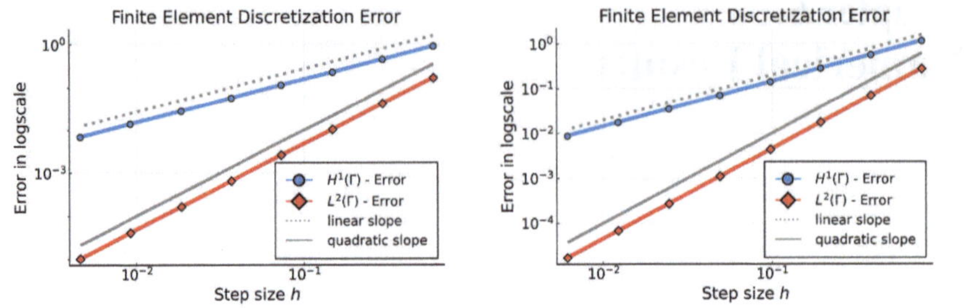

(a) (b)

Fig. 6.1 Finite element discretization error for $\mathcal{H}u + u = f$ on Γ_{star} and Γ_{dia}. *For both test problems, the* $H^1(\Gamma)$*- and* $L^2(\Gamma)$*-errors resulting from different step sizes are reported. The* $H^1(\Gamma)$*-error follows a linear slope, the* $L^2(\Gamma)$*-error a quadratic slope.* (**a**) Test Problem 2.20 (Γ_{star}). (**b**) Test Problem 2.21 (Γ_{dia})

with $\hat{\mathbf{L}}$, $\hat{\mathbf{M}}$ and $\hat{\mathbf{f}}$ as in Theorem 3.6. System (6.2) is solved for $h = 2^{-J}$, $J = 3, \ldots, 10$. The error of the finite element approximation u_h for both test problems and the various step sizes h, measured in the $H^1(\Gamma)$- and $L^2(\Gamma)$-norm, are illustrated in Fig. 6.1. Observe the quadratic decay of the $L^2(\Gamma)$- as well as a linear decay of the $H^1(\Gamma)$-error.

Let us now turn to the parabolic test problems 2.22 and 2.21. This was the heat equation

$$\frac{\partial}{\partial t}u + \mathcal{H}u = 0 \tag{6.3}$$

under Neumann-Kirchhoff conditions posed on the star and diamond graph. The initial conditions of these problems are eigenfunctions since in this case, the exact solution can be determined explicitly. A finite element semidiscretization of (6.3) with step size h leads to the system of ordinary differential equations

$$\frac{\mathrm{d}}{\mathrm{d}t}\hat{\mathbf{M}}\mathrm{u}_h(t) + \hat{\mathbf{L}}_h\mathrm{u}_h(t) = 0. \tag{6.4}$$

For the present experiment, we compute the exact solution of (6.4) at time $t = 0.1$ via the matrix exponential as

$$\mathrm{u}_h(t) = \exp(-t\,\hat{\mathbf{M}}_h^{-1}\hat{\mathbf{L}}_h)\,\mathbf{u}_h^0$$

for $h = 2^{-J}$, $J = 3, \ldots, 10$. Then, $\mathbf{u}_h^0$ denotes the discrete initial condition on the respective extended graph.

The $L^2(\Gamma)$- and $H^1(\Gamma)$-error of the finite element approximation with respect to the exact solution is evaluated for the various step sizes. We conducted the computations both with the exact mass matrix and the diagonal approximation via

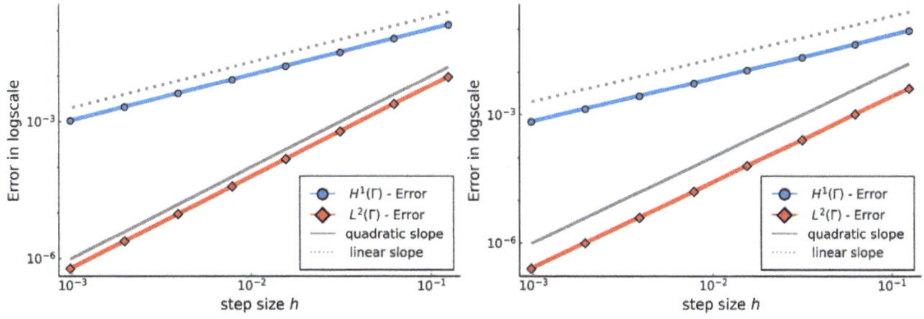

(a) (b)

Fig. 6.2 Convergence of the finite element approximation for Test Problem 2.22 (parabolic test problem on Γ_{star}). *The $H^1(\Gamma)$- and $L^2(\Gamma)$-errors resulting from different step sizes are reported, using either the exact mass matrix or a diagonal approximation. The $H^1(\Gamma)$-error follows a linear slope, the $L^2(\Gamma)$-error a quadratic slope.* (**a**) Exact mass matrix. (**b**) Mass matrix approximation

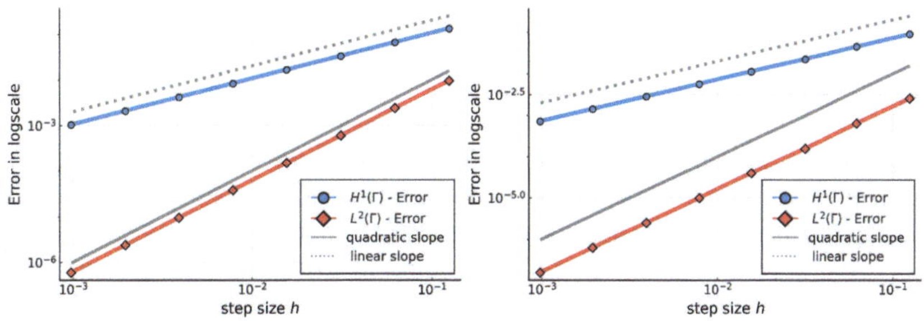

(a) (b)

Fig. 6.3 Convergence of the finite element approximation for Test Problem 2.21 (parabolic test problem on Γ_{dia}). *The $H^1(\Gamma)$- and $L^2(\Gamma)$-errors resulting from different step sizes are reported, using either the exact mass matrix or a diagonal approximation. The $H^1(\Gamma)$-error follows a linear slope, the $L^2(\Gamma)$-error a quadratic slope.* (**a**) Exact mass matrix. (**b**) Mass matrix approximation

the trapezoidal rule (compare Lemma 3.7). The results are displayed in Figs. 6.2 and 6.3. The quadratic, respectively linear, decay of the error is confirmed by the achieved results. Of course, we have repeated the experiment with several other eigenfunctions as initial conditions leading to the same convergence rates.

6.1.2 Crank-Nicolson Multigrid Method

We revisit the two parabolic test problems 2.22 and 2.21 but now solve the resulting (linear) ODE with an implicit time stepping method. To increase the order of convergence, we do not utilize the implicit Euler method as outlined in Sect. 3.3

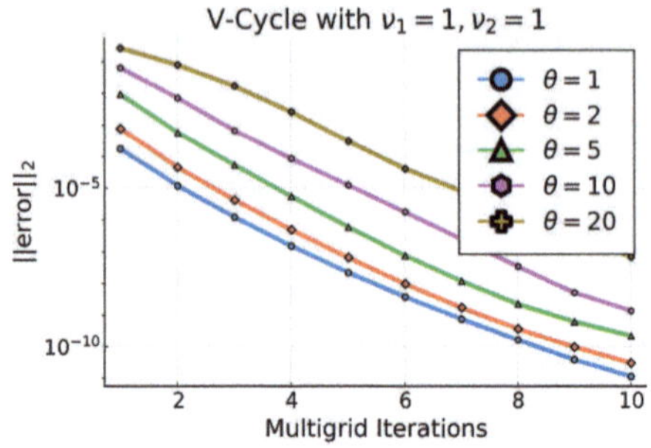

Fig. 6.4 Convergence of the multigrid algorithm for different θ. *The error of the approximate solution with respect to a solution of a direct solver is reported for the first ten multigrid iterations for different number of timesteps θ as indicated in the legend*

but apply the trapezoidal rule resulting in the iteration

$$\left(\hat{\mathbf{M}}_J + \frac{1}{2}\Delta t\,\hat{\mathbf{L}}_J\right)\mathrm{u}_J^{t+1} = \left(\hat{\mathbf{M}}_J - \frac{1}{2}\Delta t\hat{\mathbf{L}}_J\right)\mathrm{u}_J^{t}. \tag{6.5}$$

The SLE (6.5) is solved with the multigrid algorithm proposed in Sect. 3.3.2 (Algorithm 2) with coefficient matrix $\mathbf{B}_J := (\hat{\mathbf{M}}_J+\frac{1}{2}\Delta t\,\hat{\mathbf{L}}_J)$, $\mathbf{C}_J := (\hat{\mathbf{M}}_J-\frac{1}{2}\Delta t\hat{\mathbf{L}}_J)$ and $\mathbf{b}_J^t := \mathbf{0}$ (CN-MGM). As a smoother, a weighted Jacobi method is applied. We work with the discretization level $J = 10$, the coarsest level is always $J_0 = 0$.

For different numbers of time steps θ, the approximate solution $\mathrm{u}_J^\theta = \mathrm{u}_J^{0+\theta\Delta t}$ is computed by the application of the CN-MGM both using the V-cycle and W-cycle. The approximate solution u_J^θ is compared to the solution obtained by θ iterations (6.5) with a direct solver. In Fig. 6.4, the resulting error for the V-cycle is illustrated for Test Problem 2.22. The W-cycle already converged after two iterations.

The previous computations are performed for $\Delta t = h$. As pointed out by Hackbusch in [27], we observed that the convergence of the multigrid method is independent of the ratio $\Delta t/h^2$. This stands in contrast to classical smoothing iterations which can be very slow for large ratios. A comparison to a weighted Jacobi smoother applied to system (3.14) was performed and confirms this suggestion. In Fig. 6.5, we illustrate the result of the comparison of the classical Jacobi smoother to the multigrid method performed for fixed $\theta = 1$ and $J = 10$ but different step sizes Δt.

6.2 Computation of Quantum Graph Spectra

More examples applying the eigenvalue algorithm for equilateral graphs derived in Chap. 5 (Algorithm 4) as well as the generalization to non-equilateral graphs are discussed in this subsection. Moreover, the efficiency of Algorithm 4 will be investigated in comparison to a finite element discretization of a quantum graph eigenvalue problem.

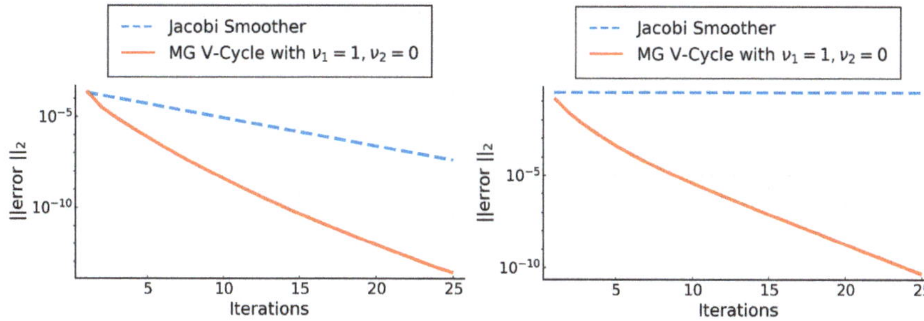

(a) (b)

Fig. 6.5 Comparison of standalone Jacobi iteration and the multigrid algorithm. *The error progressions within the first 25 iterations is compared for different step sizes* $\Delta t = h^2$ *and* $\Delta t = h^2$. (**a**) $\Delta t = h^2$. (**b**) $\Delta t = h$

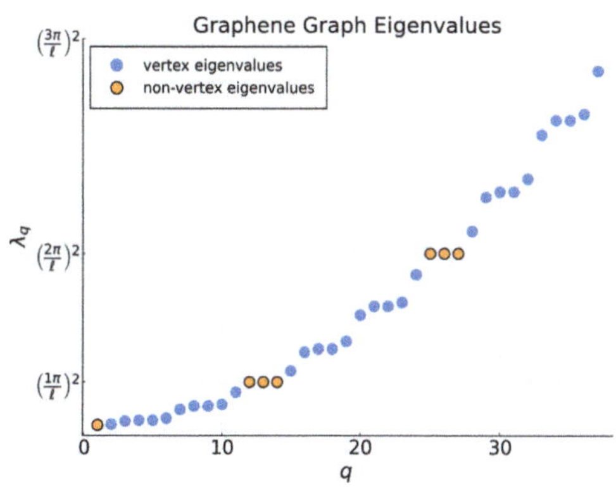

non-vertex eigenvalue $k = 0$	0
vertex eigenvalues $\sigma_{\mathcal{V},0}$	1.57e-01
	9.26e-01
	1.10e+00
	1.10e+00
	1.56e+00
	3.58e+00
	4.39e+00
	4.39e+00
	4.75e+00
	7.54e+00
non-vertex eigenvalues $k = 1$	9.87e+00
	9.87e+00
	9.87e+00
vertex eigenvalues $\sigma_{\mathcal{V},1}$	1.25e+01
	1.68e+01
	1.75e+01
	1.75e+01
	1.93e+01
	2.53e+01
	2.74e+01
	2.74e+01
	2.83e+01
	3.47e+01

Fig. 6.6 Eigenvalues of the graphene graph. *Vertex and non-vertex eigenvalues are differentiated by color in the scatter plot. For the sake of clarity, the values are also listed in the table on the right*

6.2.1 Numerical Examples for Equilateral Graphs

The following examples originate from [2] and one reason to revisit them here is to validate the eigenvalue algorithm by means of previously studied graphs. The graphs are the graphene graph Γ_{graphene} and the tree graph Γ_{tree} introduced in Example 2.25.

We start with the eigenvalues λ_q of Γ_{graphene}, which are displayed in Fig. 6.6 for $q = 1, \ldots, 50$. The first 24 eigenvalues are further reported in the table in Fig. 6.6 and they agree with the values computed in [2]. Moreover, we observe our standard patterns of vertex and non-vertex eigenvalues in the spectrum: Since the graphene graph is bipartite, it has $m - n + 2 = 3$ non-vertex eigenvalues for both k even and odd.

Interestingly, there also is a repeating pattern inside each bunch of vertex eigenvalues. Particularly, in each bunch we have two eigenvalues of multiplicity two. The corresponding eigenfunctions resolve dynamics that only occur at parts of the graph. For further examination, we illustrate the corresponding eigenfunctions in Fig. 6.7.

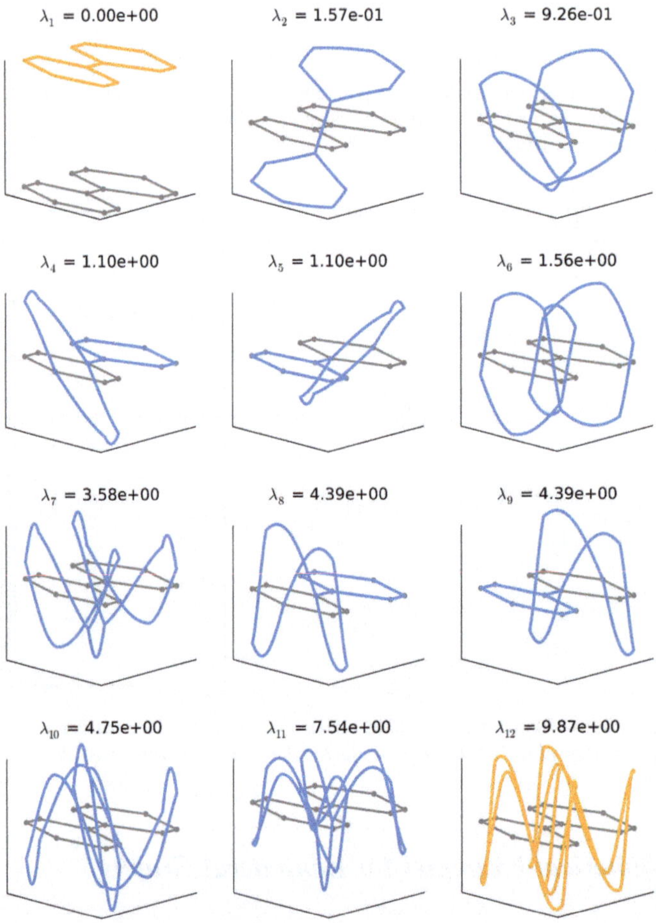

Fig. 6.7 First 12 eigenfunctions of the graphene graph. *Eigenfunctions corresponding to vertex eigenvalues are plotted in blue, the remaining in orange. The support of the eigenfunctions* ϕ_4, ϕ_5, *that correspond to the double eigenvalue* $\lambda \approx 1.10$, *contains exclusively one of the two 5-cycles of the graphene graph. The same holds true for the eigenfunctions* ϕ_8, ϕ_9 *corresponding to the second double eigenvalue* $\lambda \approx 4.39$. *Thus, the eigenvalues with multiplicity two appear in the spectrum since the eigenfunctions resolve dynamics that only occur at one of the two 5-cycles. Further, observe that* ϕ_2, ϕ_6 *and* ϕ_{10} *are all point symmetric, whereas* ϕ_3, ϕ_7 *and* ϕ_{11} *axisymmetric*

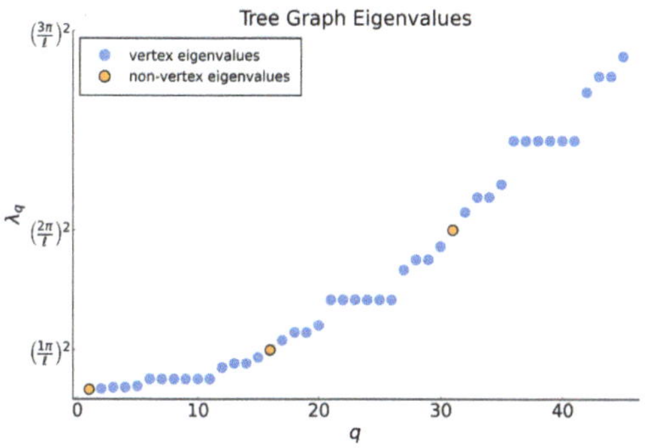

non-vertex eigenvalue $k = 0$	0
vertex eigenvalues $\sigma_{V,0}$	1.15e-01
	3.79e-01
	3.79e-01
	7.07e-01
	2.47e+00
	2.47e+00
	2.47e+00
	2.47e+00
	2.47e+00
	2.47e+00
	5.29e+00
	6.38e+00
	6.38e+00
	7.85e+00
non-vertex eigenvalues $k = 1$	9.87e+00
vertex eigenvalues $\sigma_{V,1}$	1.21e+01
	1.41e+01
	1.41e+01
	1.59e+01
	2.22e+01
	2.22e+01
	2.22e+01
	2.22e+01
	2.22e+01
	2.22e+01
	2.96e+01
	3.21e+01
	3.21e+01
	3.53e+01
non-vertex eigenvalues $k = 2$	3.95e+01

Fig. 6.8 Eigenvalues of the tree graph. *Vertex and non-vertex eigenvalues are differentiated by color in the scatter plot, eigenvalues are also listed in the table*

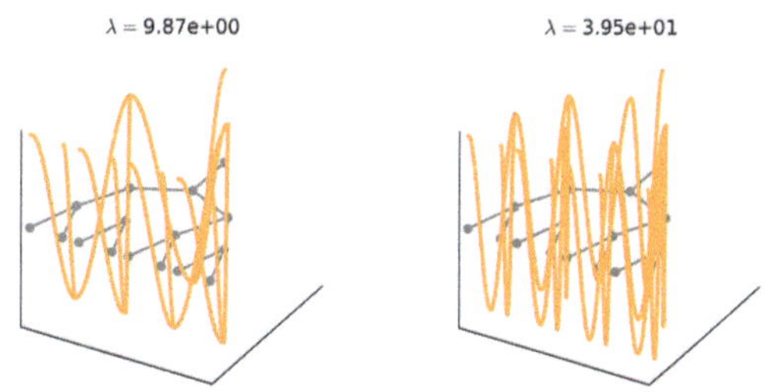

Fig. 6.9 Non-vertex eigenfunctions of the tree graph for $k = 1$ and $k = 2$. *The eigenfunctions assume either alternating values (k odd) or identical non-zero values (k even) at the vertices of the graph*

We repeat the discussion by means of the tree graph, whose eigenvalues λ_q are plotted in Fig. 6.8 for $q = 1, \ldots, 30$. Again, the computed eigenvalues are in agreement with the results in [2] as we can verify from the table in Fig. 6.8. With regard to the non-vertex eigenvalues, the tree graph is a special example (as the star graph) since $n = 16 > 15 = m$, i.e., it has more vertices than edges. Therefore, it has only one non-vertex eigenvalue for both k odd and even, although it is bipartite. The corresponding eigenfunctions are illustrated in Fig. 6.9.

As observed in the graphene graph, we can see a pattern inside each bunch of vertex eigenvalues. First, there is one eigenvalue that appears with multiplicity six ($\lambda \approx 2.47$ in the first bunch, $\lambda \approx 2.22e+01$ in the second bunch). Together with the next smaller and next larger eigenvalue, the corresponding eigenfunctions are

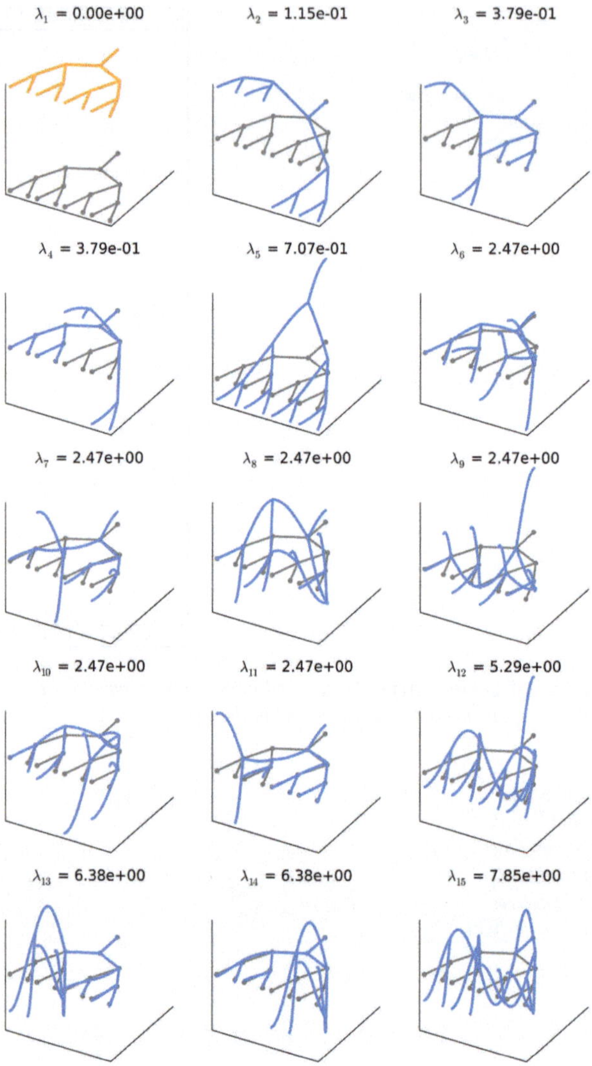

Fig. 6.10 First 15 eigenfunctions of the tree graph. *ϕ_5 and ϕ_{12} admit a symmetric behavior on the branches of the tree. In contrast, the eigenfunctions ϕ_2 and ϕ_{15} are identically zero on the stem and then behave inversely on the two main branches*

the only eigenfunctions that are not identically zero on the stem of the tree, compare Fig. 6.10.

In summary, we observed many interesting phenomenas and patterns in the spectra of the examples discussed throughout this book (note that a further numerical experiment analyzing the spectrum of lollipop graphs is conducted in Sect. 6.4.2 in the appendix). In particular, the spectrum of an equilateral metric graph is determined by the structure and connectivity patterns of the underlying combinatorial graph. This is obvious as we derived a one to one correspondence of the eigenvalues of a discrete graph Laplacian matrix to the vertex eigenvalues of the metric graph, and, for the remaining non-vertex eigenvalues, the appearance is determined by the number of vertices and edges of the graph.

6.2.2 Comparison to Finite Element Approximations

Due to the findings in Chap. 5, we are able to compute an arbitrary lower part of the spectrum of an equilateral metric graph with n vertices only using the eigenvalue decomposition of the (discrete) $n \times n$ normalized graph Laplacian matrix (Algorithm 4). Classically, it is also possible to approximatively compute eigenvalues and eigenfunctions applying a finite element discretization to the eigenvalue problem $\mathcal{H}\phi = \lambda\phi$. This leads to the generalized eigenvalue problem

$$\hat{\mathbf{L}}\mathrm{u} = \lambda\hat{\mathbf{M}}\mathrm{u} \tag{6.6}$$

with $\hat{\mathbf{L}}$ and $\hat{\mathbf{M}}$ defined as in Theorem 3.6. In [2], Arioli and Benzi discuss the finite element approximation of quantum graph spectra on the example of the tree and graphene graph from the previous subsection as well as a star graph with $n = 5$ vertices and $m = 4$ edges of length $\ell = 1$. Therefore, in this subsection, we will study accuracy and complexity of the finite element approximation in comparison to Algorithm 4 by means of these three examples. Note that the finite element approximation, in contrast to Algorithm 4, can also be applied to approximate the eigenvalue problem $\mathcal{H}\phi = \lambda\phi$ for more general operators such as $\mathcal{H} : u(x) \mapsto \frac{\mathrm{d}^2}{\mathrm{d}x^2}u(x) + \rho(x)u(x)$ with a potential $\rho(x) \in L^\infty(\Gamma)$.

Let us start with the equilateral star graph. The exact eigenvalues λ can be computed with Algorithm 4 and, for the sake of completeness, are reported in Table 6.1.

For different step sizes $h = 2^{-J}$, $J = 3, \ldots, 11$, we then solve the generalized eigenvalue problem (6.6). To streamline the exposition, we compare three of the

Table 6.1 First 9 eigenvalues of the star graph with $n = 5$ vertices

Non-vertex eigenvalue $k = 0$	0
Vertex eigenvalues $\sigma_{V,0}$	2.47e+00
	2.47e+00
	2.47e+00
Non-vertex eigenvalues $k = 1$	9.87e+00
Vertex eigenvalues $\sigma_{V,1}$	2.22e+01
	2.22e+01
	2.22e+01
Non-vertex eigenvalues $k = 2$	3.95e+01

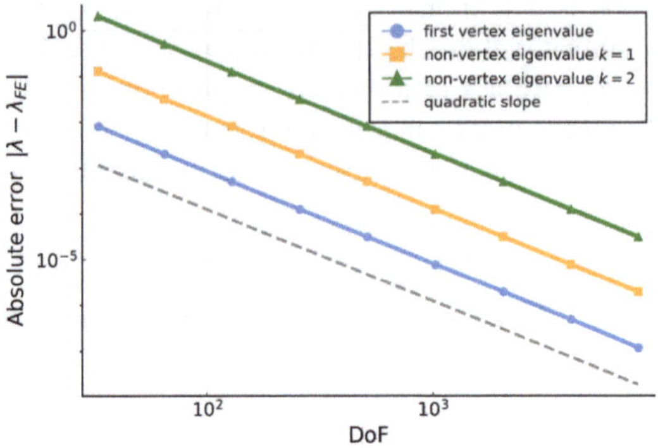

Fig. 6.11 Star graph: Convergence of finite element eigenvalue approximation. *For three selected eigenvalues, the absolute errors $|\lambda - \lambda_{\mathrm{FE}}|$ of the finite element approximation are reported for different DoF*

DoF	Absolute error		
	$\lambda_2 \approx 2.47\text{e}{+}00$	$\lambda_5 \approx 9.87\text{e}{+}00$	$\lambda_9 \approx 3.95\text{e}{+}01$
33	7.94e-03	1.27e-01	2.07e+00
65	1.98e-03	3.17e-02	5.10e-01
129	4.95e-04	7.93e-03	1.27e-01
257	1.24e-04	1.98e-03	3.17e-02
513	3.10e-05	4.95e-04	7.93e-03
1025	7.74e-06	1.24e-04	1.98e-03
2049	1.94e-06	3.10e-05	4.95e-04
4097	4.83e-07	7.74e-06	1.24e-04
8193	1.17e-07	1.93e-06	3.10e-05

approximated eigenvalues λ_{FE} to the exact values. The three selected eigenvalues are the first vertex eigenvalue (i.e., the smallest non-zero eigenvalue) as well as the non-vertex eigenvalues for k odd and k even. The absolute errors $|\lambda - \lambda_{\mathrm{FE}}|$ along with the degrees of freedom (i.e., the size of the generalized eigenvalue problem) are reported in Fig. 6.11.

For the first non-zero eigenvalue ($\lambda \approx 2.47$), a generalized eigenvalue problem of size 4097×4097 has to be solved in order to attain an accuracy of 10^{-7}. In contrast to this, the exact computation with Algorithm 4 only requires the solution of an eigenvalue problem of size $n \times n$. Remember that n is the number of vertices, which in our example is only $n = 5$. As usually in finite element eigenvalue approximations, an even smaller step size is required to approximate larger eigenvalues with a comparable accuracy. On the contrary, Algorithm 4 computes every eigenvalue of arbitrary size exactly by using the eigenvalues of the $n \times n$ normalized graph Laplacian matrix. In addition, for all eigenvalues $\lambda \neq \left(\frac{k\pi}{\ell}\right)^2$, the eigenvectors even provide a closed form representation of the corresponding eigenfunction.

In the special case of non-vertex eigenvalues $\lambda = \left(\frac{k\pi}{\ell}\right)^2$, additional degrees of freedom are necessary to compute the eigenfunctions with Algorithm 4. More precisely, an underdetermined system of linear equations of size $n + km$ needs to be solved. For the two exemplary eigenvalues in Fig. 6.11, this means we have to solve a system of size 9×9 for $\lambda \approx 9.87$ and of size 13×13 for $\lambda \approx 3.95\text{e}{+}01$. As

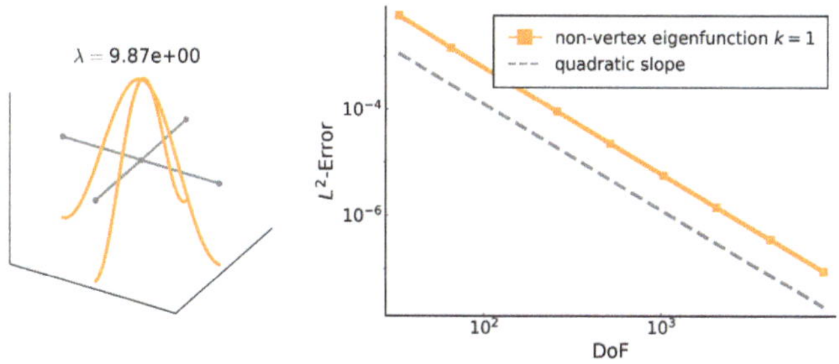

DoF	L^2-error
33	5.85e-03
65	1.44e-03
129	3.60e-04
257	8.98e-05
513	2.25e-05
1025	5.61e-06
2049	1.40e-06
4097	3.51e-07
8193	8.77e-08

(a) (b)

Fig. 6.12 Star graph: Finite element approximation of eigenfunction ϕ_5. *The $L^2(\Gamma)$-error of the finite element approximation of the eigenfunction illustrated in **(a)** is reported in **(b)** for growing DoF.* (**a**) ϕ_5 on star graph. (**b**) L^2-Error of finite element approximation

a comparison: a finite element approximation of eigenfunction ϕ_5 corresponding to $\lambda_5 \approx 9.87$ such that $\|\phi_5 - \phi_{5,\mathrm{FE}}\|_\Gamma \approx 10^{-7}$ requires $4097 = n + 1023m$ degrees of freedom, compare Fig. 6.12.

To accelerate the discussion, we briefly summarize the results of the finite element eigenvalue approximation for the graphene and tree graph. The underlying exact eigenvalues of the two graphs can be found in the previous chapter (Figs. 6.6 and 6.8). In Table 6.2, the error of the finite element approximation with different step sizes 2^{-J}, $J = 3, \ldots, 10$ is reported together with the resulting degrees of freedom for the first 12 eigenvalues.

Remember that the exact computation of the eigenvalues with Algorithm 4 requires $n = 12$ degrees of freedom for the graphene and $n = 16$ for the tree graph. In contrast, a finite element approximation of λ_{12} with $n + 1023m = 13311$ respectively $n + 1023m = 15361$ only yields an approximation of accuracy 10^{-6}.

6.2.3 Non-equilateral Graphs and the Newton-Trace Iteration

The present subsection is extracted from the joint manuscript with Chong-Son Dröge [18] with minor editorial changes. The first experiment was prepared by Chong-Son Dröge under my supervision, the second experiment was conducted and outlined by me.

As observed in the experiments in Sect. 5.3.3, the eigenvalues of the equilateral approximations converge to the exact eigenvalues for all analyzed examples. However, our aim is not to approximate the exact eigenvalues of a given non-equilateral metric graph Γ by computing arbitrary close equilateral approximations, but only to find suitable estimates to start a subsequent Newton iteration. The objective of the following experiment is to test the applicability of these eigenvalue estimates as starting values for a Newton iteration as well as the influence of the accuracy of

Table 6.2 Finite element approximation of the graphene and tree graph for the first 12 eigenvalues. *Accuracy of the finite element approximation opposed to the required degrees of freedom (size of the generalized eigenvalue problem). Note that we excluded the first eigenvalue* $\lambda_1 = 0$ *since the extended graph Laplacian* $\hat{\mathbf{L}}$ *of the generalized eigenvalue problem (6.6) always has an exact zero eigenvalue*

	Absolute error $\lvert\lambda_q - \lambda_{q,\mathrm{FE}}\rvert$ for $q = 2, \ldots, 12$										
DoF	λ_2	λ_3	λ_4	λ_5	λ_6	λ_7	λ_8	λ_9	λ_{10}	λ_{11}	λ_{12}
(a) Graphene graph											
103	3e−05	1e−03	2e−03	2e−03	3e−03	2e−02	3e−02	3e−02	3e−02	7e−02	1e−01
207	8e−06	3e−04	4e−04	4e−04	8e−04	4e−03	6e−03	6e−03	7e−03	2e−02	3e−02
415	2e−06	7e−05	1e−04	1e−04	2e−04	1e−03	2e−03	2e−03	2e−03	5e−03	8e−03
831	5e−07	2e−05	2e−05	2e−05	5e−05	3e−04	4e−04	4e−04	5e−04	1e−03	2e−03
1663	1e−07	4e−06	6e−06	6e−06	1e−05	7e−05	1e−04	1e−04	1e−04	3e−04	5e−04
3327	3e−08	1e−06	2e−06	2e−06	3e−06	2e−05	2e−05	2e−05	3e−05	7e−05	1e−04
6655	8e−09	3e−07	4e−07	4e−07	8e−07	4e−06	6e−06	6e−06	7e−06	2e−05	3e−05
13311	1e−09	7e−08	1e−07	1e−07	2e−07	1e−06	2e−06	2e−06	2e−06	5e−06	8e−06
(b) Tree graph											
121	2e−05	2e−04	2e−04	7e−04	8e−03	8e−03	8e−03	8e−03	8e−03	8e−03	4e−02
241	4e−06	5e−05	5e−05	2e−04	2e−03	2e−03	2e−03	2e−03	2e−03	2e−03	9e−03
481	1e−06	1e−05	1e−05	4e−05	5e−04	5e−04	5e−04	5e−04	5e−04	5e−04	2e−03
961	3e−07	3e−06	3e−06	1e−05	1e−04	1e−04	1e−04	1e−04	1e−04	1e−04	6e−04
1921	7e−08	7e−07	7e−07	3e−06	3e−05	3e−05	3e−05	3e−05	3e−05	3e−05	1e−04
3841	2e−08	2e−07	2e−07	6e−07	8e−06	8e−06	8e−06	8e−06	8e−06	8e−06	4e−05
7681	4e−09	5e−08	5e−08	2e−07	2e−06	2e−06	2e−06	2e−06	2e−06	2e−06	9e−06
15361	5e−10	1e−08	1e−08	4e−08	5e−07	5e−07	5e−07	5e−07	5e−07	5e−07	2e−06

the estimate on the number of required Newton iterations. As in Example 5.29, the experiments will be conducted to compute the first positive eigenvalue λ_2.

We will consider two different sample graphs, each with random edge lengths $\ell_e \in [1, 2]$, rounded to two decimal digits. These include a star graph with $n = 6$ vertices and $m = 5$ edges and a Barabási-Albert graph with $n = 50$ vertices and $m = 96$ edges. As the edge lengths are rounded to two decimal digits, the exact eigenvalues can be computed through an equilateral representation with edge length $h = 0.01$.

Since we have observed (Fig. 5.18) that the exact eigenvalues lie between the eigenvalues of the floor and ceil approximation, we use the mean of the eigenvalues as the initial guess for the Newton-trace iteration, i.e.,

$$\lambda^{\text{init}} := \frac{\lambda_2(\mathfrak{G}_{\text{fl},h}) + \lambda_2(\mathfrak{G}_{\text{cl},h})}{2}. \tag{6.7}$$

Here, $\lambda_2(\mathfrak{G}_{\text{fl},h})$ denotes the first positive eigenvalue of the equilateral floor approximation and $\lambda_2(\mathfrak{G}_{\text{cl},h})$ of the equilateral ceil approximation. In the following experiment, we now want to investigate how many Newton iterations are required when starting with λ^{init} as defined in (6.7) for various $h = 2^{-J}$, $J = 2, \ldots, 6$.

As a stopping criterion for the Newton-trace iteration, we monitor the reciprocal condition number of $\mathbf{H}(z)$ (see (5.2)) defined as $\kappa^{-1}(\mathbf{H}) := \frac{1}{\|\mathbf{H}\| \, \|\mathbf{H}^{-1}\|}$. Since $\mathbf{H}(z)$ is singular at the eigenvalues, a reciprocal condition number close to zero indicates that the Newton-trace iteration converged to an eigenvalue λ. In the following experiment, we stop the Newton-trace iteration whenever $\kappa^{-1}(\mathbf{H}) < 10^{-10}$.

In Table 6.3, we have recorded the exact eigenvalue λ_2, the initial value λ^{init} arising from the equilateral approximations with length h as well as the number of required Newton-trace iterations until convergence to λ_{NEP}. First of all, it is important to notice that for each initial value λ^{init}, the Newton-trace iteration converges to the same eigenvalue λ_{NEP}. Given that the $\mathbf{H}(z)$-matrix in the Newton-trace iteration is only of size $n \times n$, the number of required iterations is moderate, even for the "worst" choice of λ^{init}. As expected from the results in the previous subsections, a decreasing edge length h leads to better initial values, which in turn further reduces the number of required Newton-trace iterations.

Table 6.3 Number of Newton-trace iterations (N_{iter}) required for computing λ_2, for the star and the Barabási-Albert graph [18]

	Star, $\lambda_2 \approx 0.701372$			Barabási-Albert, $\lambda_2 \approx 0.212386$		
h	λ^{init}	N_{iter}	λ_{NEP}	λ^{init}	N_{iter}	λ_{NEP}
0.25	0.785305	8	0.701372	0.221519	11	0.212386
0.125	0.725887	4	0.701372	0.213850	4	0.212386
0.0625	0.710570	3	0.701372	0.211420	4	0.212386
0.03125	0.698711	3	0.701372	0.212036	3	0.212386
0.015625	0.704270	3	0.701372	0.212253	3	0.212386

So far, we have only studied the convergence towards the smallest non-zero eigenvalue. The question is whether further eigenvalues can still be found reliably by equilateral approximations, i.e., if patterns in the spectrum can be reflected sufficiently. In other words, how can we guarantee that all eigenvalues are found? Moreover, for the previous experiments, we have examined example graphs of moderate size and with a limited number of decimal digits for the edge lengths. This guaranteed that the eigenvalue problem of the equilateral approximations can be easily solved by standard eigenvalue algorithms. Other phenomena might occur when considering more complex, large graphs. In general, for small step sizes h, this results in large scale linear eigenvalue problems. In Sect. 5.3.3, we therefore proposed to follow a *nested iteration approach* (Algorithm 7).

With this nested iteration eigenvalue solver, we will finally consider a random Barabási-Albert graph with $n = 500$ vertices and $m = 1491$ edges with a randomly assigned edge length $\ell_e \in [1, 5]$. For $h = 2^{-J}$, $J = 1, \ldots, 8$, we each time compute the $Q = 10$ smallest eigenvalues of equilateral floor and ceil approximations $\mathfrak{G}_{\mathrm{cl},h}$, $\mathfrak{G}_{\mathrm{fl},h}$ and run the Newton-trace iteration with initial value $\lambda_q^{\mathrm{init}} = \frac{1}{2}\left(\lambda_q(\mathfrak{G}_{\mathrm{fl},h}) + \lambda_q(\mathfrak{G}_{\mathrm{cl},h})\right)$. In this scenario, it is not feasible to compute the exact solution using equilateral representations. Instead, we plot the reciprocal condition number of $\mathbf{H}(z)$, the roots of which indicate the eigenvalues as illustrated in the upper plot in Fig. 6.13.

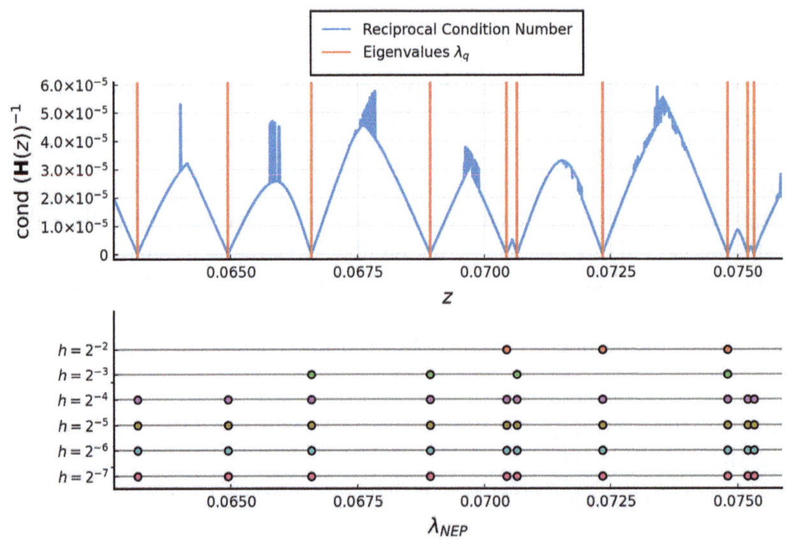

Fig. 6.13 Reciprocal condition number of $\mathbf{H}(z)$ for Barabási-Albert graph with $n = 500$ vertices compared to computed eigenvalues by equilateral approximations with $\mathfrak{G}_{\mathrm{cl},h}$, $\mathfrak{G}_{\mathrm{fl},h}$ and a subsequent Newton-trace iteration [18]. *Observe that the spacing between the roots is very irregular. In the lower part of Fig. 6.13, the eigenvalues found by the Newton iteration with* $\lambda_i^{\mathrm{init}}$ *are plotted for the different levels* $J = 1, \ldots, 8$

Table 6.4 Number of required Newton-trace iterations until convergence to λ_q of the Barabási-Albert graph with $n = 500$ vertices [18]

	λ_1	λ_2	λ_3	λ_4	λ_5	λ_6	λ_7	λ_8	λ_9	λ_{10}
$h = 2^{-2}$	–	–	–	–	11	–	9	18	–	–
$h = 2^{-3}$	–	–	10	16	–	3	–	24	–	–
$h = 2^{-4}$	4	2	3	3	6	3	4	3	4	3
$h = 2^{-5}$	4	3	3	3	3	3	4	3	3	2
$h = 2^{-6}$	2	2	3	3	2	2	3	2	3	2
$h = 2^{-7}$	2	2	2	2	2	2	1	2	2	2

Clearly, a large step size is not sufficient to resolve all patterns in the spectrum, as not all eigenvalues can be found. In particular, we observed that this effect appears with large, sparse graphs. As a criterion to estimate the quality of the equilateral approximations, we suggest to verify that $\lambda_q(\mathfrak{G}_{\text{fl},h}) > \lambda_q(\mathfrak{G}_{\text{cl},h})$ for $q = 1, \ldots, Q$. Finally, even for large graphs, the number of required Newton iterations remains moderate, as shown in Table 6.4.

6.3 Spectral Solution Method

In this subsection, we study numerical results in the context of the spectral solution method. The decay of coefficients, spectral convergence of the truncation error, and more aspects concerning the projection coefficients are investigated. We verify the Weyl's law type eigenvalue estimates and finally analyze the spectral solution of the generalized heat and the reaction-diffusion equation.

6.3.1 Decay of Coefficients

The objective of this subsection is to numerically investigate the approximation of a given function u on Γ as linear combination of eigenfunctions, i.e., by

$$u_Q = \sum_{q \leq Q} c_q \phi_q. \tag{6.8}$$

In Sect. 4.1.2, we derived that the decay of coefficients in the spectral expansion u_Q significantly determines the truncation error. For $u \in \text{dom}_{\mathcal{H},\text{NK}}$ sufficiently smooth such that all even derivatives are in $\text{dom}_{\mathcal{H},\text{NK}}$, the projection coefficients are bounded by

$$|c_q| \leq \frac{1}{\lambda_q^d} \left\| u^{(2d)} \right\|_\Gamma \tag{6.9}$$

leading to the truncation error

$$\left\|u - P_Q u\right\|_\Gamma \le \frac{1}{\lambda_Q^d}\left\|u^{(2d)}\right\|_\Gamma,$$

compare Theorem 4.4 and Corollary 4.5. Recall that in [10], this spectral convergence has been only proven for functions with compact support on an edge of the graph, whereas we extended the class of applicable functions.

As a first example, we consider a function u on a small star graph.

Example 6.1 Consider the star graph Γ_{star} with $n = 5$ vertices and $n = 4$ edges, each of length 1 and let u be a function with

$$u_{e'} = \exp\left(-\frac{(x - \ell/2)^2}{(\ell/10)^2}\right)$$

for a fixed edge $e' \in \mathcal{E}$ and $u_e(x) = 0$ for all $e \neq e'$.

The choice of u in Example 6.3 originates from [10]. We adapted this choice since it is a nice example of a smooth function with compact support on one edge that cannot be trivially resolved by the eigenfunctions. Being of compact support on one edge, it will be later interesting to study heat evolution with an initial condition posed on one edge only.

For the numerical treatment, it is irrelevant on which edge the non-zero part of u is defined, we could for example choose u as illustrated in Fig. 6.14a. For this function, the projection coefficients c_q of the spectral expansion (6.8) are computed for $q = 1, \ldots, 120$. The results are plotted in Fig. 6.14b along with the upper bound resulting from the choice of $d = 3$ in (6.9). Observe the spectral decay of the projection coefficients and that the upper bound is sharp for every q.

For different choices of Q, we moreover computed the truncation error and again visualize it together with the error bound for $d = 3$ in Fig. 6.14c.

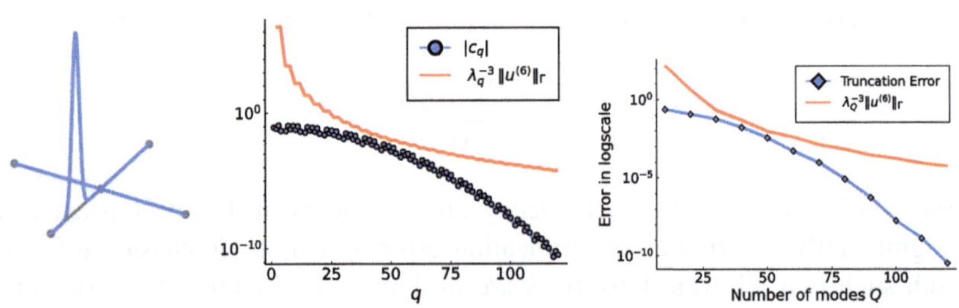

(a) (b) (c)

Fig. 6.14 Decay of coefficients and truncation error for Test Problem 6.1. *The chosen initial condition is illustrated in* (**a**)*. The values of the first 120 projection coefficients are plotted in* (**b**)*, the truncation error for Q number of modes in* (**c**)*. Note that we have suppressed the coefficients with* $|c_q| = 0$ *to clarify the illustration.* (**a**) u on Γ. (**b**) Decay of coefficients. (**c**) Truncation error

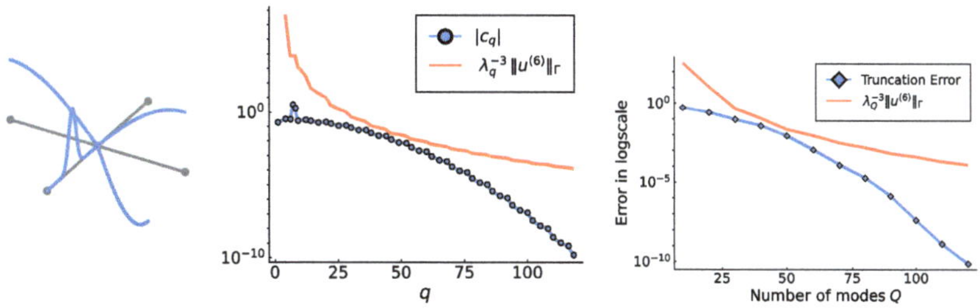

(a) (b) (c)

Fig. 6.15 Decay of coefficients and truncation error for Test Problem 6.2. *The chosen initial condition is illustrated in (**a**). The values of the first 120 projection coefficients are plotted in (**b**), the truncation error for Q number of modes in (**c**). Note that we have suppressed the coefficients with* $|c_q| = 0$ *to clarify the illustration.* (**a**) u on Γ. (**b**) Decay of coefficients. (**c**) Truncation error

In fact, the function in Example 6.1 has compact support on one edge and thus fits in the category of functions for which spectral accuracy has been shown in [10]. We therefore repeat the experiment with a function meeting the requirements in Theorem 4.4 without compact support.

Example 6.2 Consider again the star graph Γ_{star} from Example 6.1 but with equilateral edge length $\ell = \pi/2$ and let

$$u_{e_1}(x) = -2\sin(x), \quad u_{e_2} = \sin(x), \quad u_{e_3} = \sin(x), \quad u_{e_4} = \exp\left(-\frac{(x-\ell/2)^2}{(\ell/10)^2}\right).$$

The function introduced in Example 6.2 is plotted in Fig. 6.15 together with the projection coefficients as well as the truncation error. The results are visually identical to those of the previous example, with the only exception that two additional eigenfunctions ($q = 2, 3$) are explored in the spectral expansion. In contrast, the corresponding projection coefficients have been identical zero in Example 6.1.

As indicated, we have so far suppressed the projection coefficients that has been identical zero already for small q in the plots. The associated eigenfunctions are not relevant for the representation of the function on the graph. To this end, some interesting phenomena have been observed during numerical experiments which we will demonstrate by means of examples in Sect. 6.4.2 in the appendix.

6.3.2 *Eigenvalue Estimates from Weyl's Law*

In Sect. 4.1.3, we have seen that the number of eigenvalues smaller than a given value can be bounded from below and above by

$$\frac{\text{vol}_\Gamma}{\pi}\lambda_q - m \leq |\{\lambda \in \sigma(\Gamma) : \lambda \leq \lambda_q^2| \leq \frac{\text{vol}_\Gamma}{\pi}\lambda_q + n.$$

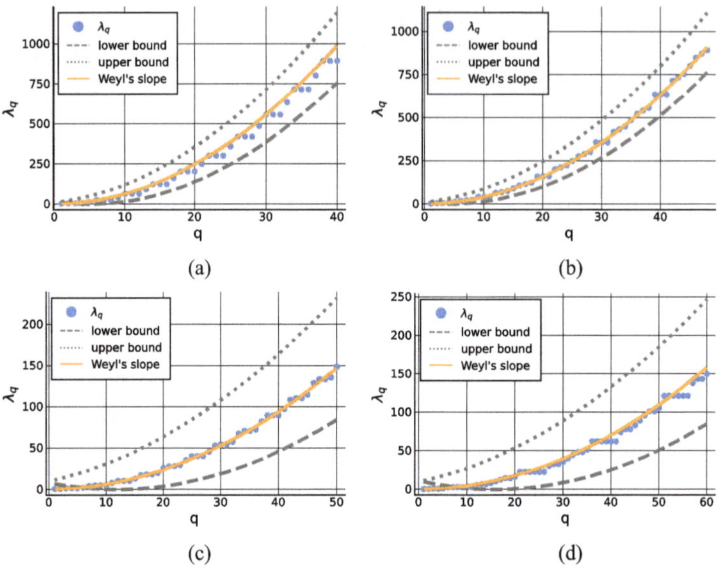

Fig. 6.16 Eigenvalue estimates from Weyl's law. *Exact eigenvalues are plotted as dots and closely follow Weyl's slope printed as indicated in the legend. Upper and lower bounds are plotted as dashed lines and are sharp for q large enough.* (**a**) Star graph. (**b**) Diamond graph. (**c**) Graphene graph. (**d**) Tree graph

In particular, for q sufficiently large, the q-th eigenvalue λ_q satisfies

$$\left(\frac{(q-n)\pi}{\mathrm{vol}_\Gamma}\right)^2 \leq \lambda_q \leq \left(\frac{(q+m)\pi}{\mathrm{vol}_\Gamma}\right)^2.$$

We verify the eigenvalue estimates by means of the following example graphs: our common star and diamond graph as well as the graphene and tree graph. For all graphs, we choose the equilateral edge length $\ell = 1$. The results are reported in Fig. 6.16 and support the derived estimates. As pointed out in [10], one observes that the asymptotic behavior of the eigenvalues can be described by *Weyl's slope*, i.e.,

$$\lambda_q \sim \left(\frac{q\pi}{\mathrm{vol}_\Gamma}\right)^2. \tag{6.10}$$

However, if we repeat the experiment for a random Barabási-Albert and Erdős-Rényi graph with $n = 100$ vertices and parameters $d = 3$ and $p = 0.1$ respectively, (6.10) merely describes an improved lower bound, compare Fig. 6.17. This effect occurs when $m \gg n$ due to the large patterns of non-vertex eigenvalues $\left(\frac{k\pi}{\ell}\right)^2$ for each $k \in \mathbb{N}$. In contrast, if $m \approx n$ (as in the previous examples), the eigenvalues follow Weyl's slope much better.

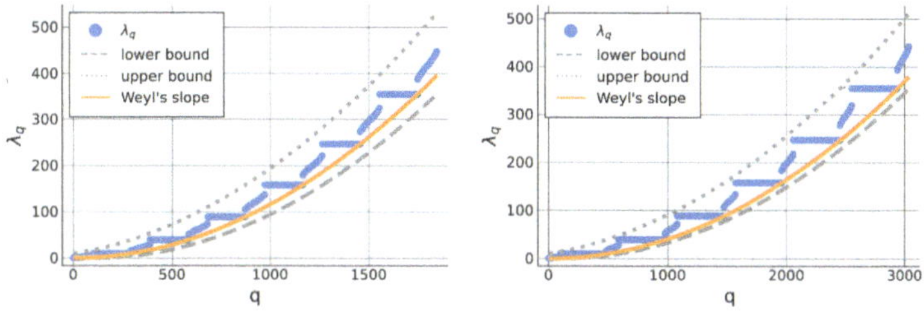

(a) (b)

Fig. 6.17 Eigenvalue estimates from Weyl's law for random graphs. *Again, eigenvalues are plotted as dots together with Weyl's slope. Observe that Weyl's slope is close to the lower bound indicated as dashed line.* (**a**) Barabási-Albert graph. (**b**) Erdős-Rényi graph

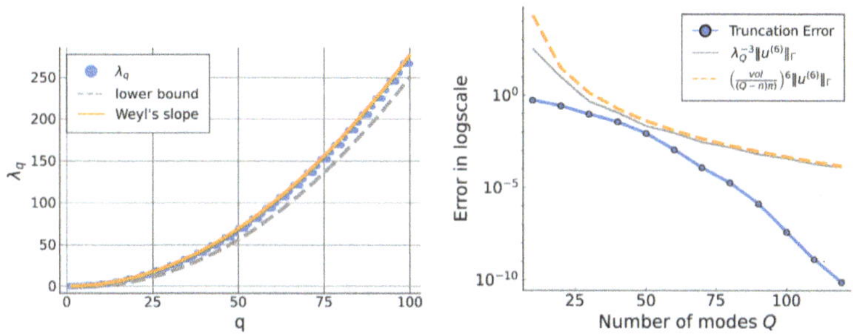

(a) (b)

Fig. 6.18 Eigenvalue estimates and truncation error for Example 6.2. *Eigenvalue estimates for Γ_{star} by Weyl's law are illustrated in (**a**) and applied to compute an upper bound for the truncation error in (**b**).* (**a**) Eigenvalue estimates. (**b**) Truncation error

The purpose of our interest in eigenvalue estimates was to derive the estimate

$$\left\| u - P_Q u \right\|_\Gamma \leq \left(\frac{\text{vol}_\Gamma}{(Q-n)\pi} \right)^{2d} \left\| u^{(2d)} \right\|_\Gamma$$

for the truncation error, see Corollary 4.7. For the star graph and u from Example 6.2, we therefore repeat the experiment from the previous section and, in Fig. 6.18, report the truncation error together with the upper bound for $d = 3$.

6.3.3 Heat Equation

We analyze the spectral solution of the classical heat equation $\frac{\partial u}{\partial t} + \mathcal{H}u = 0$ on Γ. The corresponding semidiscretized initial value problem

$$\frac{\mathrm{d}}{\mathrm{d}t}\mathbf{c}(t) + \mathbf{\Lambda}\mathbf{c}(t) = \mathbf{0}, \quad \mathbf{c}(0) = \mathbf{c}_0,$$

can be readily solved by

$$\mathbf{c}(t) = \exp(-t\boldsymbol{\Lambda})\,\mathbf{c}^0.$$

In this expression, $\boldsymbol{\Lambda}$ is a diagonal matrix of eigenvalues and the coefficients $\mathbf{c}^0$ are given by orthogonal projections of the initial condition u^0, compare Theorem 4.8. Thus, the solution at time t is given by

$$\boldsymbol{u}_Q(t) = \left(\exp(-t\lambda_1)\left(u^0, \phi_1\right)_\Gamma\right)\phi_1 + \ldots + \left(\exp(-t\lambda_Q)\left(u^0, \phi_Q\right)_\Gamma\right)\phi_Q. \tag{6.11}$$

We already showed that $\lambda_1 = 0$ and all the remaining eigenvalues λ_q are positive. For $t \to \infty$, the solution therefore (as expected) tends to the equilibrium state

$$\lim_{t\to\infty} \boldsymbol{u}_Q(t) = \left(u^0, \phi_1\right)_\Gamma \phi_1$$

where ϕ_1 is constant (compare Theorem 5.10). As a result, the $L^2(\Gamma)$-norm approaches

$$\lim_{t\to\infty} \left\|\boldsymbol{u}_Q(t)\right\|_\Gamma = \left(u^0, \phi_1\right)_\Gamma \tag{6.12}$$

and the $H^1(\Gamma)$-seminorm converges to $\lim_{t\to\infty} |\boldsymbol{u}_Q(t)|_{H^1(\Gamma)} = \left(u^0, \phi_1\right)_\Gamma \frac{\mathrm{d}\phi_1}{\mathrm{d}x} = 0$.

We verify the reasoning by examining the behavior of the solution of the heat equation on a tree graph, as well as its $L^2(\Gamma)$-norm and $H^1(\Gamma)$-seminorm.

Example 6.3 Let Γ_{tree} be the tree graph from Example 2.25 with $\ell = 1$. In this example, we consider the heat equation on Γ_{tree} given the initial condition

$$u_e(x) = \exp\left(-\frac{(x-\ell/2)^2}{(\ell/10)^2}\right)$$

on the stem of the tree and $u_{e'}(x) = 0$ on all remaining edges e'.

It is important to choose an initial condition that fulfills the Neumann-Kirchhoff conditions. In the finite element experiments, we have therefore chosen eigenfunctions. This choice is not reasonable here since we can resolve this condition with one basis function.

We initially consider a short time interval $[0, 0.1]$ and plot the initial condition $(t = 0)$ as well as the solution $\boldsymbol{u}_Q(t)$, $Q = 225$ for $t = 0.025, 0.5, 0.1$ in Fig. 6.19b. The $L^2(\Gamma)$-norm of the initial condition can be computed in closed form and is given by $\left\|u^0\right\|_\Gamma = 0.354022$. By (6.12), we expect the norm to approach $\lim_{t\to\infty} \left\|u_Q(t)\right\|_\Gamma = \left(u^0, \phi_1\right)_\Gamma$ where ϕ_1 is a normalized, constant eigenfunction

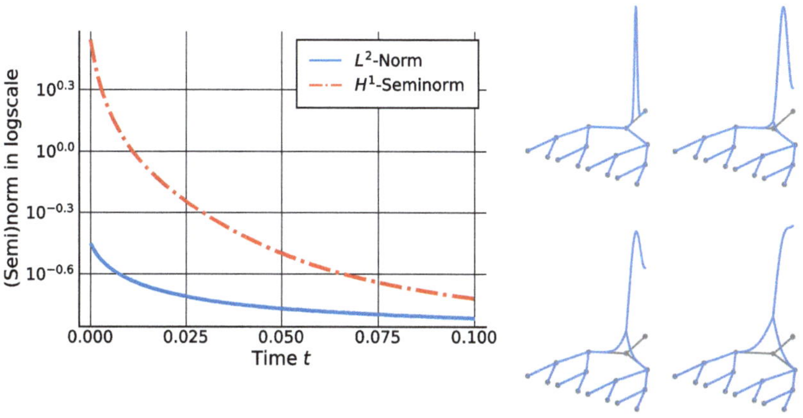

(a) (b)

Fig. 6.19 Heat equation on tree graph. (**a**) L^2-norm and H^1-seminorm of $\boldsymbol{u}_Q(t)$ for $t \in [0, 0.1]$. (**b**) Solution at $t = 0, 0.025, 0.05, 0.1$

Fig. 6.20 Heat equation on tree graph: L^2-norm and H^1-seminorm of $\boldsymbol{u}_Q(t)$ for $t \in [0, 4]$

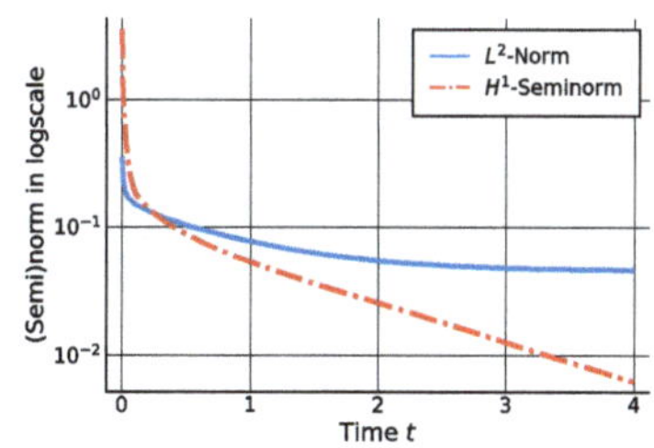

on Γ, i.e., $\phi_1 \equiv \sqrt{1/m}$. Thus, $\left(u^0, \phi_1\right)_\Gamma = 0.0457646$, which agrees with the observed decrease of the $L^2(\Gamma)$-norm in Fig. 6.19a.

Moreover, we can confirm the convergence in the long run by repeating the experiment for the time interval $[0, 4]$. As depicted in Fig. 6.20, the $L^2(\Gamma)$-norm converges to 0.0457646 and the $H^1(\Gamma)$-seminorm vanishes for $t \to \infty$.

Let us now investigate the convergence of the solution in dependence on the graph structure. A similar experiment was considered in [2] where the authors revealed that the decay behavior of the norms depends on the eigenvalues of the graph, which, in turn, are widely studied as connectivity measure in combinatorial graph theory. Here, we modified the experiment in that we compare three graphs with a similar number of edges. This is important as we found out that the $L^2(\Gamma)$-norm of the solution tends to $\left(u^0, \phi_1\right)_\Gamma$ with $\phi_1 \equiv \sqrt{1/m}$. If we choose the same initial condition on each graph (as before we are going to pose it on a single edge), this ensures that the solution converges to an equilibrium in the same order of magnitude, which allows comparing the decay behavior.

Example 6.4 Let Γ_{BA} be a Barabási-Albert graph with $n = 9, d = 2$ and $\Gamma_{\text{tree}}, \Gamma_{\text{graphene}}$ the tree and graphene graph from Example 2.25, each with equi-

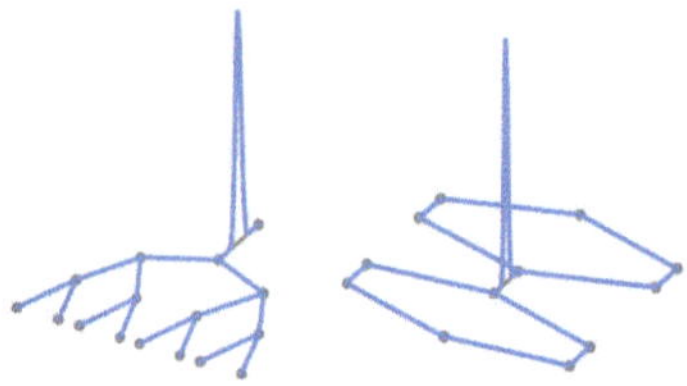

(a)

	Γ_{tree}	Γ_{graphene}	Γ_{BA}
λ_1	0.00e+00	0.00e+00	0.00e+00
λ_2	1.15e-01	1.57e-01	1.27e+00
λ_3	3.79e-01	9.26e-01	1.58e+00
λ_4	3.79e-01	1.10e+00	1.81e+00
λ_5	7.07e-01	1.10e+00	2.47e+00

(b)

Fig. 6.21 Initial conditions and eigenvalues for Example 6.4. (**a**) Initial condition on Γ_{tree} and Γ_{graphene}. (**b**) Eigenvalues λ_q for $q = 1, \ldots, 6$

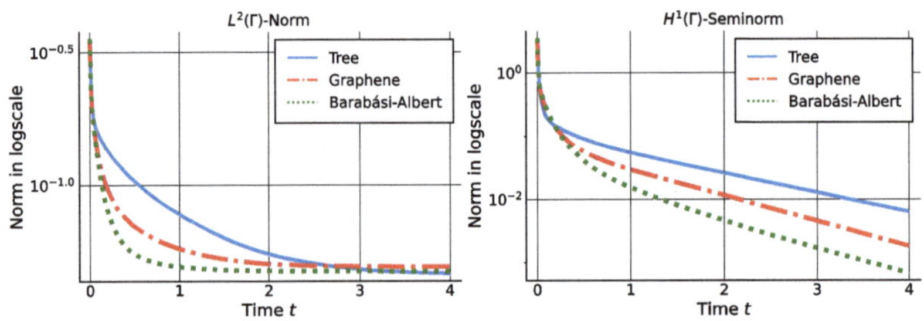

Fig. 6.22 Heat equation on tree, graphene and Barabási-Albert graph (Example 6.4). $L^2(\Gamma)$*-norm and* $H^1(\Gamma)$*-seminorm of* $\boldsymbol{u}_Q(t)$, $t \in [0, 4]$ *for the three example graphs*

lateral edge length $\ell = 1$. The initial condition of the IBVP for the heat equation is given by

$$u_e(x) = \exp\left(-\frac{(x - \ell/2)^2}{(\ell/10)^2}\right)$$

on one edge $e \in \mathcal{E}$ and $u_{e'}(x) = 0$ on all remaining edges $e' \neq e$.

The initial condition on the tree and graphene graph are illustrated in Fig. 6.21a. For the Barabási-Albert graph, we choose a random edge e with $u_e \neq 0$. The first five eigenvalues of the graphs are recorded in Fig. 6.21b. As pointed out in [2], the small eigenvalues mainly determine the convergence of the solution. This is obvious from (6.11) since $\exp(-t\lambda)$ vanishes more slowly for smaller λ. In other words, the larger the second eigenvalue λ_2, the faster the expected convergence of the $L^2(\Gamma)$-Norm.

This anticipated behavior is supported by our numerical findings. In Fig. 6.22, we can see that the $L^2(\Gamma)$-norm approaches the equilibrium fastest for the Barabási-Albert graph and slowest for the tree graph, which is in agreement with the ordering of the eigenvalues λ_2. The decay of the $H^1(\Gamma)$-seminorm behaves in the same manner.

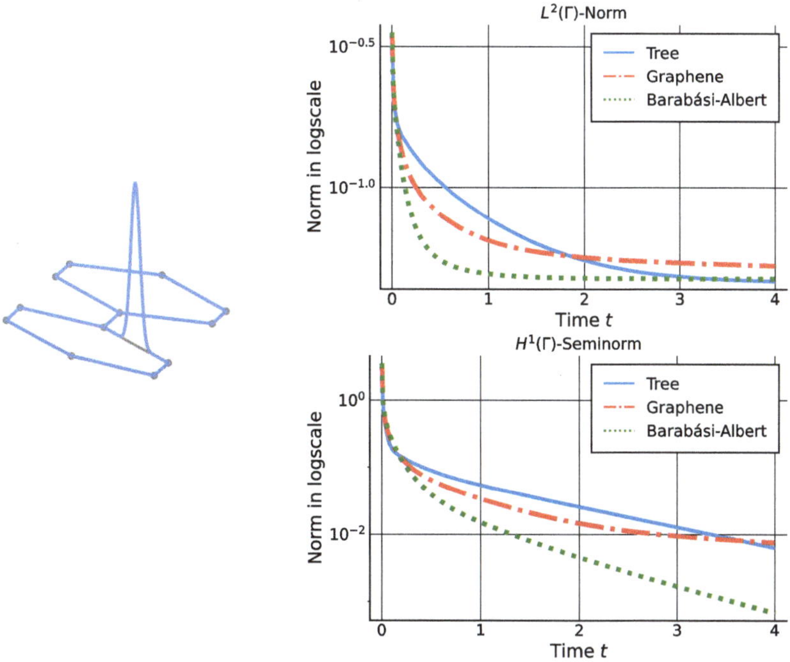

Fig. 6.23 $L^2(\Gamma)$-norm and $H^1(\Gamma)$-seminorm of $\boldsymbol{u}_Q(t)$ for $t \in [0, 4]$. *Here,* $\boldsymbol{u}_Q$ *is the solution of the heat equation on the tree and Barabási-Albert graph from Example 6.4 and the graphene graph with initial condition as illustrated in the left figure*

However, there is another important factor: the initial condition. This can be seen very clearly by means of the tree graph. The eigenfunction associated with the smallest non-zero eigenvalue $\lambda_2 \approx 1.15\text{e}{-}01$ is identically zero on the stem of the tree (compare Fig. 6.10 from the previous section), i.e., on the only edge where the initial condition $u^0 \neq 0$. This implies that the coefficient of ϕ_2 in the spectral expansion (6.11) is zero and thus the convergence of the solution can not be influenced by λ_2 at all. The reason why the tree graph still showed the slowest convergence in the previous experiment is that the same holds true for the eigenfunction ϕ_2 of Γ_{graphene}. If we repeat the experiment with the initial condition of the graphene graph posed on a different edge, the results emerge confirming to the previous observations as we can see in Fig. 6.23. In fact, although the decay for the graphene graph is faster than for the tree graph at the beginning, it decelerates in the long term such that the equilibrium is reached much later.

6.3.4 Fractional Diffusion

The framework of the spectral Galerkin method allows to naturally deal with a fractional Laplacian operator without further effort. As indicated in Sect. 4.4, the fractional diffusion equation

$$\frac{\partial}{\partial t}u - \Delta^{\alpha} u = 0, \quad \alpha \in \mathbb{R}$$

on Γ leads to the spectral Galerkin semidiscretization

$$\frac{\mathrm{d}}{\mathrm{d}t}\mathbf{c}(t) + \mathbf{\Lambda}^{\alpha}\mathbf{c}(t) = 0.$$

We briefly investigate some observed particularities of fractional heat conduction on metric graphs. Let us first consider the diamond graph Γ_{dia} with initial condition

$$u_e(x) = \exp\left(-\frac{(x-\ell/2)^2}{(\ell/10)^2}\right)$$

on an arbitrary chosen edge e and $u_{e'}(x) = 0$ otherwise. The solution $\boldsymbol{u}_Q(t)$, $Q = 148$ obtained with $\alpha = 1, 0.5, 0.1$ at different time points is illustrated in Fig. 6.24. Observe the increasing non-local effects for small α.

We moreover compute the $L^2(\Gamma)$-norm and $H^1(\Gamma)$-seminorm for $t \in [0, 2]$ and compare the norms for the different choices of α in Fig. 6.25. Both norms decay slowest for $\alpha = 0.1$ and fastest for $\alpha = 1$. Since $\lambda > 1$ for all eigenvalues of Γ_{dia}, this observation is in line with the reflections from the previous section.

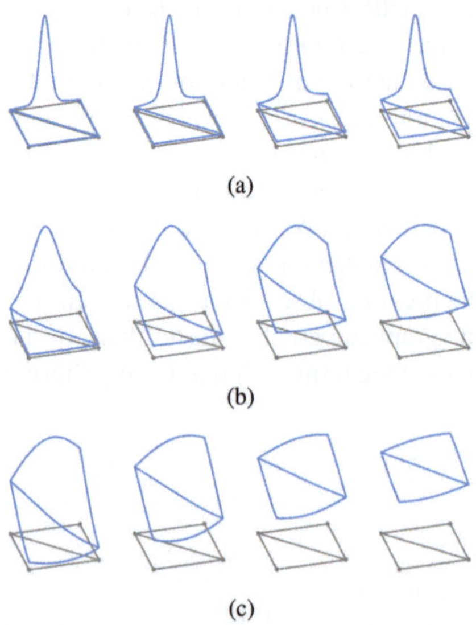

Fig. 6.24 Solution $\boldsymbol{u}_Q$ of the fractional diffusion equation on Γ_{dia}. *Plotted for* $\alpha = 0.1, 0.5, 1$ *and for* $t = 0.25, 0.5, 0.75, 1$ *(from left to right)*. **(a)** $\alpha = 0.1$. **(b)** $\alpha = 0.5$. **(c)** $\alpha = 1$

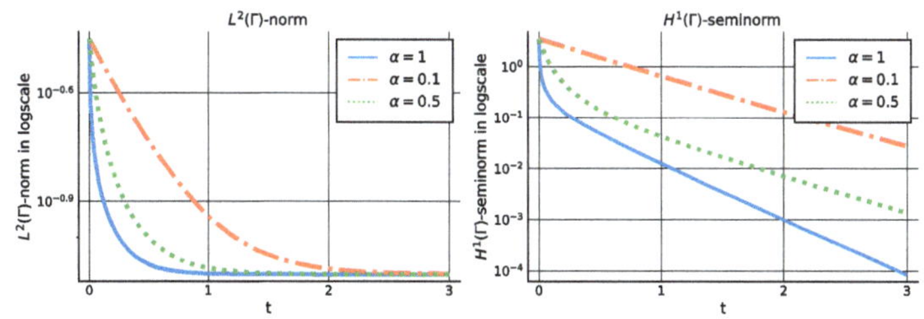

Fig. 6.25 $L^2(\Gamma)$-norm and $H^1(\Gamma)$-seminorm of $\boldsymbol{u}_Q(t)$ for $t \in [0, 3]$.*Here,* $\boldsymbol{u}_Q$ *is the solution of the fractional diffusion equation on* Γ_{dia} *with* α *as indicated in the legend*

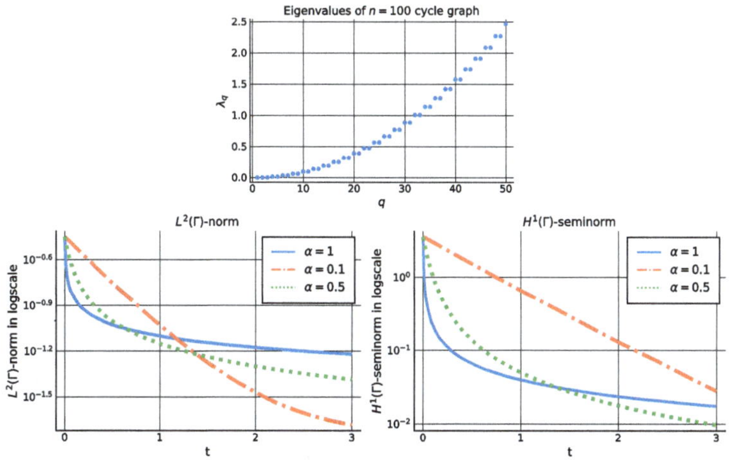

Fig. 6.26 $L^2(\Gamma)$-norm and $H^1(\Gamma)$-seminorm of $\boldsymbol{u}_Q(t)$ for $t \in [0, 3]$. *Here,* $\boldsymbol{u}_Q(t)$, *is the solution of the fractional diffusion equation on a cycle graph with* $n = 100$ *vertices and* α *as indicated in the legend*

However, the effect is reversed when considering graphs with very small eigenvalues. As an extreme example, we report the behavior of the norms for a large cycle graph with $n = 100$ vertices in Fig. 6.26 together with its first eigenvalues. Observe that the cycle graph has a large number of eigenvalues $\lambda \ll 1$ explaining the fast decay of the $L^2(\Gamma)$-norm for $\alpha = 0.1$.

6.3.5 Reaction-Diffusion Equations

We conclude this section with experiments concerning (semi)linear reaction-diffusion equations

$$\frac{\partial u}{\partial t} + \mathcal{H}u = \mathcal{R}(u)$$

subject to Neumann-Kirchhoff conditions and with given initial condition u^0. We will start by a comparison of the behavior of the solution for linear reaction terms

$$\mathcal{R}_1(u) \equiv 1, \ \mathcal{R}_2(u) = u$$

and nonlinear reaction terms

$$\mathcal{R}_3(u)) = u^{\frac{1}{2}}, \ \mathcal{R}_4(u) = u^{1.5} \text{ and } \mathcal{R}_5(u) = u^2.$$

Let us again consider the tree graph along with the following initial condition.

Example 6.5 Let Γ_{tree} be the tree graph from Example 2.25 with equilateral edge length $\ell = 1$. As initial condition, we choose a linear combination of two eigenfunctions, namely

$$u^0(x) = \phi_1(x) + \phi_5(x),$$

as illustrated in Fig. 6.27.

First, we pose the reaction-term only on one edge of the tree, the stem, and set it to zero otherwise. We compute the spectral solution $\boldsymbol{u}_Q(t)$, $Q = 300$ for $t \in [0, 1]$ as well as its $L^2(\Gamma)$-norm. As can be seen in Fig. 6.28, the $L^2(\Gamma)$-norm grows fastest at the beginning for the constant reaction term, however, in the long term, the quadratic term outruns and the norms grow in the expected order: $\mathcal{R}_5, \mathcal{R}_4, \mathcal{R}_2, \mathcal{R}_3, \mathcal{R}_1$.

We repeat the experiment with the reaction term posed on all edges of the graph. Again, the $L^2(\Gamma)$-norm exhibits the strongest growth for the quadratic reaction term, compare Fig. 6.29. Moreover, we computed the $H^1(\Gamma)$-seminorm and plotted its course for $\mathcal{R}_1, \mathcal{R}_2, \mathcal{R}_3$. For the remaining reaction terms, the $H^1(\Gamma)$-seminorm does not decrease.

The next experiment seeks to analyze the convergence of the spectral Galerkin method. Again, we consider the reaction-diffusion equation with the previously introduced reaction terms but the star graph along with the initial condition of

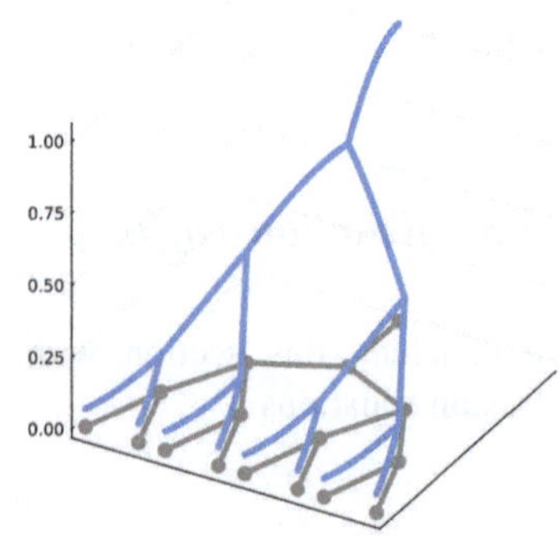

Fig. 6.27 Initial condition $u^0(x) = \phi_1(x) + \phi_5(x)$ on the tree graph

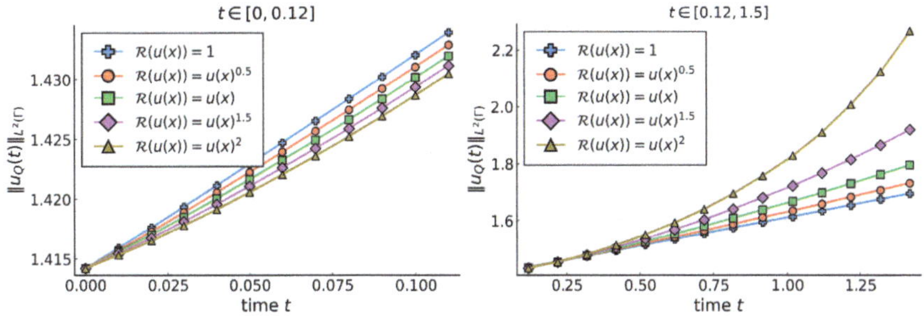

Fig. 6.28 Reaction-diffusion equation on a tree graph—$L^2(\Gamma)$-norm of the spectral Galerkin approximation. *The progression of the* $L^2(\Gamma)$*-norm over time is compared for different reactions terms* $\mathcal{R}$*, all of them posed on a single edge only (the stem)*

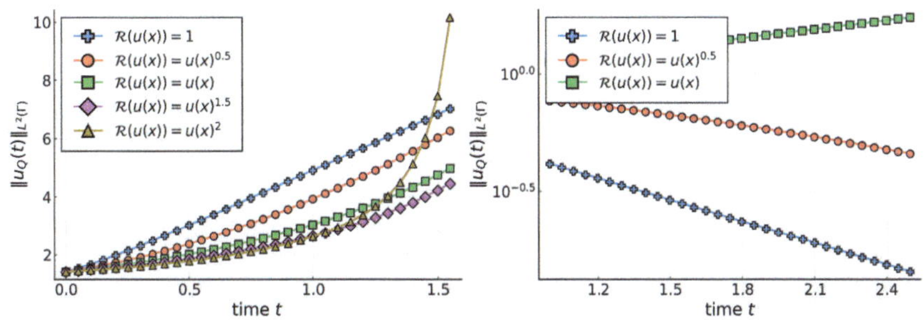

(a) (b)

Fig. 6.29 Reaction-diffusion equation on a tree graph—$L^2(\Gamma)$-norm and $H^1(\Gamma)$-seminorm of the spectral Galerkin approximation. *The norms are compared over time for different reactions terms* $\mathcal{R}$*, each posed on every edge.* (**a**) Progression of the $L^2(\Gamma)$-norm. (**b**) Progression of the $H^1(\Gamma)$-seminorm

Example 6.1. First, we compute a reference solution $\boldsymbol{u}_{\text{ref}}(t)$ with $Q_{\text{ref}} = 120$ modes at $t = 10^{-8}$ by applying the exponential Euler iteration with $\Delta t = 10^{-8}$. We then compute the spectral solution for different $Q < Q_{\text{ref}}$ along with the error with respect to the reference solution, i.e., $\left\|\boldsymbol{u}_{\text{ref}}(t) - \boldsymbol{u}_Q(t)\right\|_\Gamma$. The differences in the results for the multiple reaction terms are marginal, but for the sake of completeness, we included the output in Fig. 6.30.

We further examine the development of the error over time for Example 6.1 with the various reaction terms as above and a reference solution $\boldsymbol{u}_{\text{ref}}$ with $Q_{\text{ref}} = 120$. We then compute the spectral Galerkin approximation $\boldsymbol{u}_Q(t)$ with $Q = 90$ for the time interval $[0, 10^{-5}]$. To compute the solutions $u_{\text{ref}}(t)$ and $\boldsymbol{u}_Q(t)$, we again the exponential Euler method with $\Delta t = 10^{-7}$ and compute $\left\|\boldsymbol{u}_{\text{ref}}(t) - \boldsymbol{u}_Q(t)\right\|_\Gamma$ for each timepoint. We compare the error evolution for the different reaction terms, illustrated in Fig. 6.31.

For $\mathcal{R}_1$, we are in the case of the generalized heat equation where we expect the error not to grow over time. This agrees with our numerical results in Fig. 6.31.

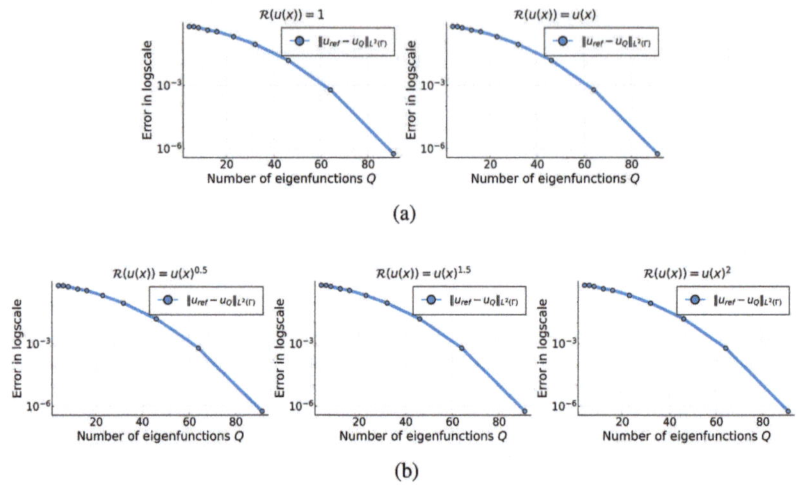

Fig. 6.30 Reaction-diffusion equation for Example 6.1. *Convergence of the spectral solution* $\boldsymbol{u}_Q$ *to a reference solution* ($Q_{\text{ref}} = 120$) *for different reaction terms* $\mathcal{R}$. (**a**) Linear reaction terms. (**b**) Nonlinear reaction terms

Fig. 6.31 Reaction-diffusion equation on a star graph: Evolution of the spectral approximation error over time. *The error of the spectral Galerkin approximation* $\boldsymbol{u}_Q$ *with* $Q = 90$ *is illustrated for different reaction terms and* $t \in [2^{-6}, 3^{-6}]$. *Graph and initial condition as in Example 6.1*

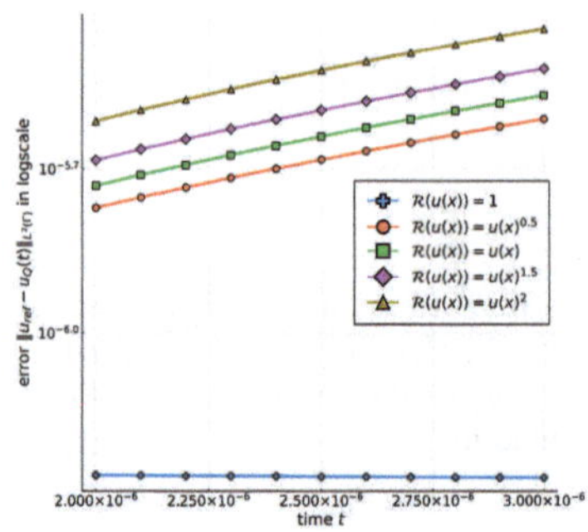

In the nonlinear setting, we expect the error to grow with time, which again is in agreement with the evolution of the error in the previous example. Moreover, we observe that the growth depends on the reaction term, or, to be more precise, on its Lipschitz-constant (note that local Lipschitz-continuity often suffices on a bounded time interval).

6.4 Comparison of Finite Element and Spectral Galerkin Method

One question we have to raise after presenting the two Galerkin discretizations is how the achieved accuracy is related to the complexity of the method. The spectral convergence of the spectral Galerkin method means that the error decreases much faster compared to the finite element method when additional basis functions are applied. The number of utilized basis functions (i.e., the dimension of the trial space) in turn determines the degrees of freedom of the (semi)discretizations and thereby the computational costs. However, the computational complexity of the methods depends on the considered PDE and a universal comparison is not straightforward.

6.4.1 Elliptic Test Problem

We want to discuss and compare the convergence and computational costs by means of a simple, elliptic test problem. Let us therefore consider the following example whose exact solution can be specified.

Test Problem 6.13 *Let Γ_{star} be a star graph with $n = 5$ vertices and $m = 4$ edges of length $\ell = 1$. Let further*

$$u_e = \exp\left(\frac{-(x - x_0)^2}{s^2}\right)$$

with $x_0 = \ell/2$, $s = \ell/12$ on one fixed edge e and $u_{e'} = 0$ on the remaining edges e'. Then, $u \in \mathrm{dom}_{\mathcal{H},\mathrm{NK}}$ is the solution of

$$\mathcal{H}u + u = f \tag{6.13}$$

with

$$f_e = -\frac{(\exp(-(x - x_0)^2/s^2)(4(x - x_0)^2 - 2s^2)}{s^4} + \exp\left(-\frac{(x - x_0)^2}{s^2}\right)$$

for e and $f_{e'} = 0$ for $e' \neq e$.

An approximate solution of (6.13) with the given right-hand side is computed with both the finite element and the spectral Galerkin approximation. In the finite element setting, the discretization leads to the system of linear equations

$$\hat{\mathbf{L}}\mathrm{u} + \hat{\mathbf{M}}\mathrm{u} = \hat{\mathbf{f}} \tag{6.14}$$

with $\hat{\mathbf{L}}$, $\hat{\mathbf{M}}$ and $\hat{\mathbf{f}}$ as in Theorem 3.6. For a given step size $h = 2^{-J}$, $J \in \mathbb{N}$, the degrees of freedom are given by $n + m(2^J - 1)$. The system is assembled and solved for different step sizes $h = 2^{-J}$, $J = 1, \dots, 14$, and the approximation error, measured in the $L^2(\Gamma)$-norm, is computed.

On the other hand, the spectral Galerkin approximation of (6.13) leads to the discretized system

$$\mathbf{\Lambda c} + \mathbf{c} = \mathbf{f} \tag{6.15}$$

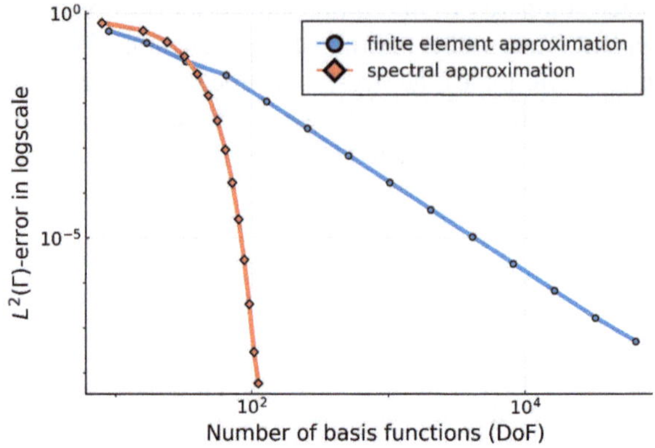

Fig. 6.32 Convergence of the finite element and spectral Galerkin discretizations of Test Problem 6.13. *The number of applied basis functions are plotted against the* $L^2()$*-error*

with $\boldsymbol{\Lambda}$ and $\mathbf{f}$ as in Theorem 4.8. The degrees of freedom are given by the number of eigenfunctions Q used to represent the solution u_Q, i.e., the dimension of the trial space. We compute the discretization for $Q = 7, \ldots, 100$ and again evaluate the $L^2(\Gamma)$-error of the spectral approximation. The results are presented in Fig. 6.32, where the degrees of freedom are contrasted to the achieved accuracy. One can clearly see the spectral convergence of the spectral method, whereas the error of the finite element approximation improves quadratically.

To achieve the best accuracy possible, we applied a direct solver to solve the finite element discretization (6.14) in the previous results. Moreover, the integrals in the right-hand sides for both the finite element and spectral discretization have been computed with a standard quadrature implementation in Julia (`QuadGK`).

For the following discussion of the computational costs, we solve the finite element discretization with a nested iteration multigrid algorithm instead. The multigrid algorithm works similarly to the one presented in Sect. 3.3.2 but is now applied to the system (6.14) instead of (3.14). For each $J = 1, \ldots, 14$, we follow a nested iteration strategy (see for example [28]) to compute a start vector for the multigrid iteration at level J. The latter in turn is ran with a V-Cycle, coarsest level $J_0 = 0$, and $\nu_1 = \nu_2 = 1$ smoothing steps. The iteration is stopped when the residual is in the order of magnitude of the discretization error. The nested iteration multigrid is implemented matrix free, i.e., only the application of the discretization matrices as well as intergrid operators are computed, compare Sect. 3.3.4. Explicit matrices only have to be assembled at the coarsest level to apply a direct solver. These costs are negligible here since the coarsest system is of size 5×5. Moreover, the integrals in the right-hand side $\hat{\mathbf{f}}$ are approximated by an application of the mass matrix $\hat{\mathbf{M}}$ to the discrete function.

With these configurations, the time to solve (6.14) for different step sizes $J = 1, \ldots, 14$ is recorded using the package `BenchmarkTools` in Julia (version 1.8.2). The computations were performed on an 2021 iMac with Apple M1 chip (8-core CPU with 4 performance cores and 4 efficiency cores, 7-core GPU, 16-core Neural Engine, 16 GB unified memory). The elapsed computation time in dependence of the required degrees of freedom is depicted in Fig. 6.33a. The observed linear slope

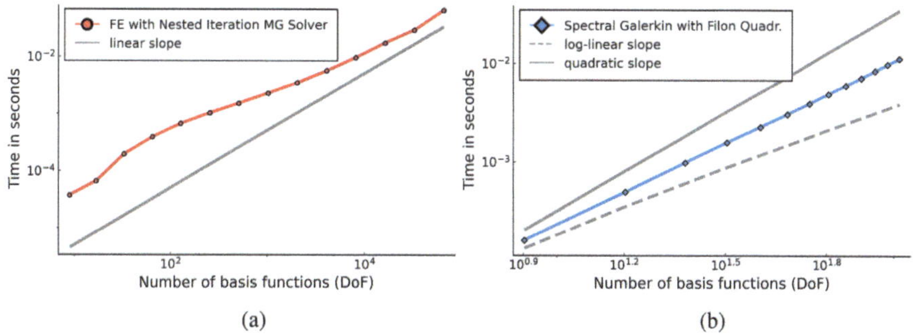

Fig. 6.33 Complexity of the finite element and spectral Galerkin discretizations of Test Problem 6.13.*The elapsed computation time is plotted versus the number of applied basis functions.* (**a**) Finite element approximation. (**b**) Spectral Galerkin approximation

is in agreement with the expected complexity of a classical nested iteration multigrid method.

For the spectral discretization, we compute the integrals in the right-hand side of (6.15) with the Filon-type quadrature rule derived in Sect. 4.3.2. The number of quadrature nodes applied is chosen such that the integrals are approximated up to an error in the order of the expected discretization error. The baseline eigenvalue problem required for the computation of the vertex eigenvalues and eigenfunctions is solved with a direct, standard Julia method (`eigen`) (applied to the symmetric normalized graph Laplacian instead of the harmonic graph Laplacian as outlined in Sect. 5.2.3). For the non-vertex eigenvalues, a sparse eigenvalue solver (`eigs`) is applied to compute the eigenvectors. The computation time grows less than quadratically in terms of the number of basis functions, see Fig. 6.33b.

Finally, we compare the achieved accuracy in dependence of the computational time for both methods in Fig. 6.34. The experiment demonstrates the efficiency of the spectral method if highly accurate solutions are required. To be more precise, the spectral method outperforms the finite element method if an $L^2(\Gamma)$-error less than $1\text{e}{-}05$ is attempted.

More general elliptic problems and problems with lower regularity (e.g. $u \in H^1(\Gamma)$) should be discussed in future work.

6.4.2 Parabolic Problems

The discussion is less straightforward for parabolic problems since the computational costs depend on several factors. Let us for instance consider the heat equation

$$\frac{\partial}{\partial t}u + \mathcal{H}u = 0 \tag{6.16}$$

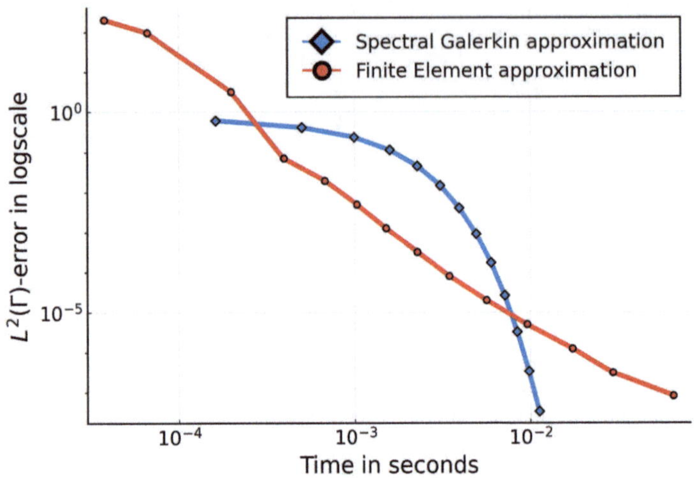

Fig. 6.34 Computation time versus achieved accuracy of the finite element and spectral Galerkin approximation for Test Problem 6.13

on the star graph from Test Problem 6.13 with initial condition

$$u_{e'}^0 = \exp\left(-\frac{(x-\ell/2)^2}{(\ell/10)^2}\right)$$

for a fixed edge $e' \in \mathcal{E}$ and $u_e^0(x) = 0$ for all $e \neq e'$. A finite element semidiscretization of the heat equation (6.16) was derived in Theorem 3.6 and is given by

$$\frac{\mathrm{d}}{\mathrm{d}t}\hat{\mathbf{M}}\mathrm{u}(t) + \hat{\mathbf{L}}\mathrm{u}(t) = 0, \quad \mathrm{u}(0) = \mathbf{u}^0. \tag{6.17}$$

If we apply the Crank-Nicolson method to approximate the solution at time T, in total $T/\Delta t$ SLEs need to be solved. In contrast, the spectral Galerkin discretization of (6.16),

$$\frac{\mathrm{d}}{\mathrm{d}t}\mathbf{c}(t) + \boldsymbol{\Lambda}\mathbf{c}(t) = 0, \quad \mathbf{c}(0) = \mathbf{c}^0 \tag{6.18}$$

(compare Theorem 4.8), can be solved directly for arbitrary T since $\boldsymbol{\Lambda}$ is diagonal. Namely, the coefficients of the solution are given by $\mathbf{c}(T) = \exp(-T\boldsymbol{\Lambda})\mathbf{c}^0$. Instead, the computational costs are dominated by the computation of the eigenfunction basis and the projection of the initial condition. Thus, they behave like the cost of solving an elliptic problem as discussed earlier.

The computational time for the solution at different time points $T = 0.001, \ldots, 0.2$ is measured for a finite element discretization with step size

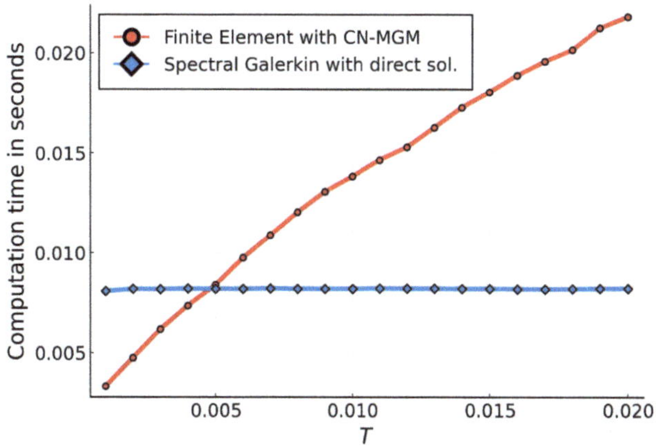

Fig. 6.35 Computation time of the finite element and spectral Galerkin approximation of (6.16) for different time points T

$h = 2^{-10}$ (i.e., 4097 DoF) and a spectral Galerkin discretization with $Q = 88$ modes. According to the experiments for Test Problem 6.13, this choice of basis functions is expected to result in a solution with comparable accuracy. The SLEs arising in the Crank-Nicolson method are solved with a multigrid method as outlined in Sect. 3.3.2 (CN-MGM solver).

The results are illustrated in Fig. 6.35 and confirm that the computation time for the spectral solution is independent of T whereas it grows with T for the finite element solution. In the particular example, the spectral method outperforms the finite element approach in terms of computational time for $T \geq 0.005$.

Appendix

Further Examples for the Computation of Quantum Graph Spectra

Consider the following two only slightly different *lollipop graphs* the first one is a 3-cycle graph attached to a path consisting of one edge, the second example is a 3-cycle graph attached to a path consisting of two edges, compare Fig. 6.36. Both graphs are not bipartite and have equilateral edge length $\ell = 1$.

The objective of this example is a comparison of their eigenvalues and eigenfunctions, or, in other words, to investigate the effect of the additional edge on patterns in the spectrum. Let us begin with the eigenvalues, illustrated in Fig. 6.37 for $\lambda < \left(\frac{3\pi}{\ell}\right)^2$.

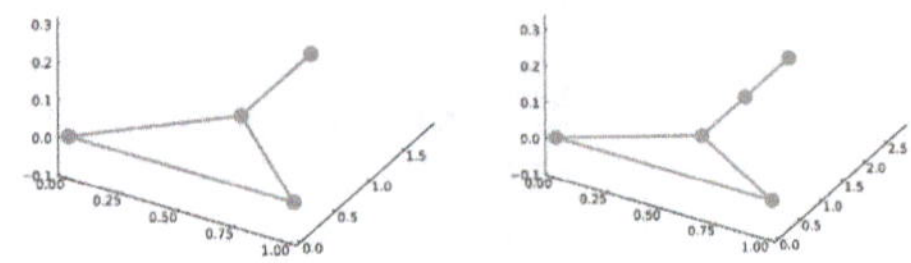

Fig. 6.36 Lollipop graphs

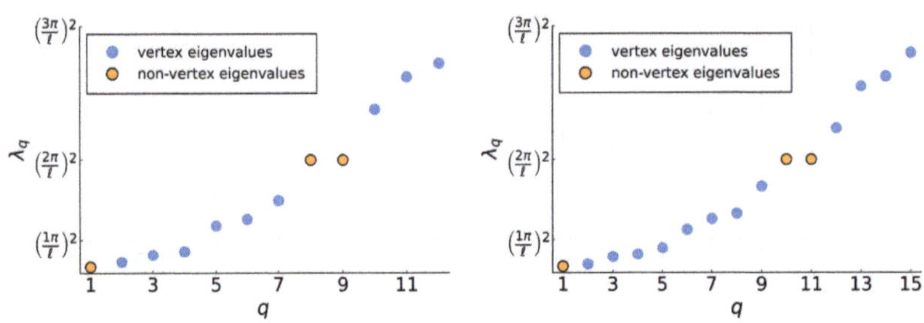

(a) (b)

Fig. 6.37 Eigenvalues of lollipop graphs. (**a**) Lollipop graph 1. (**b**) Lollipop graph 2

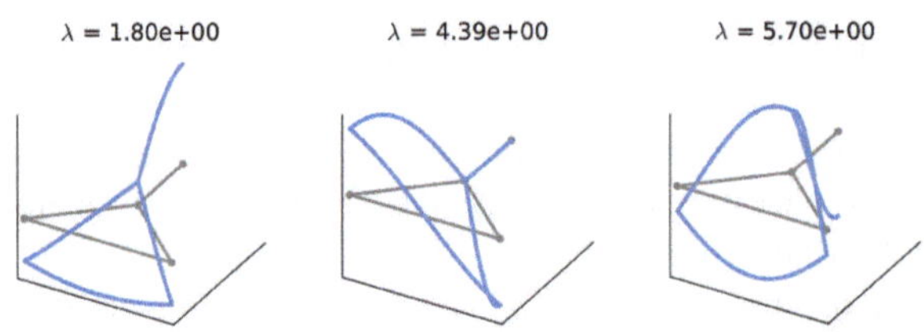

(a)

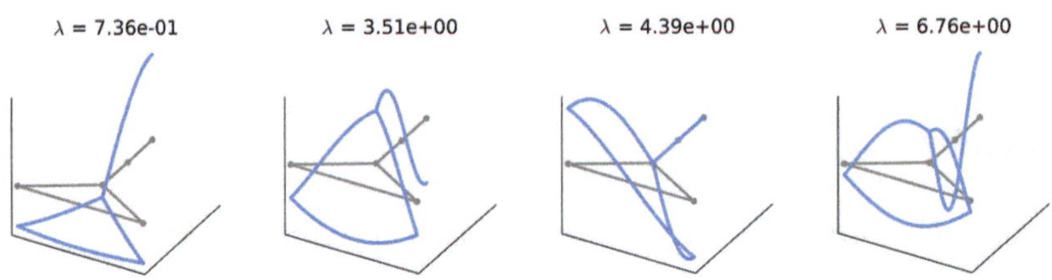

(b)

Fig. 6.38 First bunch of vertex eigenfunctions for lollipop graphs. (**a**) Lollipop graph 1. (**b**) Lollipop graph 2

The number of non-vertex eigenvalues matches for both graphs: $m - n + 2 = 2$ for k even and $m - n = 0$ for k odd. Each vertex bunch of lollipop graph 1 has three eigenvalues, for lollipop graph 2 we have four as it has an additional vertex. To examine the role of the additional vertex eigenvalue, we take a closer look at the first bunch of vertex eigenfunctions plotted in Fig. 6.38.

One eigenvalue ($\lambda = 4.39$) appears in both spectra. The corresponding eigenfunction describes the same dynamic on both graphs. In particular, it is identical zero on the "stem". Similarly, the eigenfunction of lollipop graph 1 corresponding to $\lambda = 1.80$ has an equivalent in lollipop graph 2 describing a similar dynamic (ϕ_2).

Note that the spectrum of lollipop graph 2 is equivalent to a lollipop graph of the first type, but with a stem of length $\ell_e = 2$ since the elimination of the vertex does not change the spectrum.

Analysis of Projection Coefficients in the Spectral Expansion

As indicated in Sect. 6.3.1, we suppressed the projection coefficients identical to zero already for small q in the results. In this context, some interesting phenomena that have been observed during numerical experiments will be exemplary demonstrated on the following experiment.

Example 6.6 Consider the diamond graph Γ_{dia} with $n = 4$ vertices and $n = 5$ edges, each of length 1 and let u be a function with

$$u_{e'} = \exp\left(-\frac{(x - \ell/2)^2}{(\ell/10)^2}\right)$$

for a fixed edge $e' \in \mathcal{E}$ and $u_e(x) = 0$ for all $e \neq e'$. In particular, we choose $e' = e_3$ as the middle edge of the diamond graph such that u assumes the form illustrated in Fig. 6.39b.

Let us for convenience return to the notation we used in deriving the equilateral graph spectra: for each eigenvalue μ of the harmonic graph Laplacian matrix, we have $\lambda_{\mu,k}$, $k = 0, 1, 2, \ldots$ eigenfunctions of the metric graph. In total, this gives us $n - 1$ different types of vertex eigenfunctions (if Γ is not bipartite, otherwise we have $n - 2$). Each of them describes a fixed dynamic on the vertices of the graph and for increasing k, solely the oscillation across the edges increases (compare the derivation and examples in Sect. 5.2.1). These observations are also reflected in the projection coefficients. To be more precise, in Fig. 6.39a, we illustrate the coefficients c_q for the eigenfunctions corresponding to the vertex eigenvalues $\lambda_{\mu,k}$, $k = 0, 1, \ldots, 6$. In fact, $c_q \neq 0$ only for all eigenvalues associated with the harmonic graph Laplacian eigenvalues μ_2 and μ_3.

A deeper examination of the eigenfunctions immediately explains this behavior. Consider for instance ϕ_λ for $\lambda = \lambda_{\mu_2,k}$, $k = 0, 1$ in Fig. 6.39c. Obviously, the projection coefficient $c_\lambda = \int_\Gamma u(x)\phi_\lambda(x)\mathrm{d}x = 0$ since $\phi_\lambda \equiv 0$ on e_3 (the middle edge). For μ_3, the integral vanishes since u is axisymmetric on edge e_3, whereas the eigenfunctions corresponding to $\lambda_{\mu_3,k}$ are pointsymmetric, compare Fig. 6.39d.

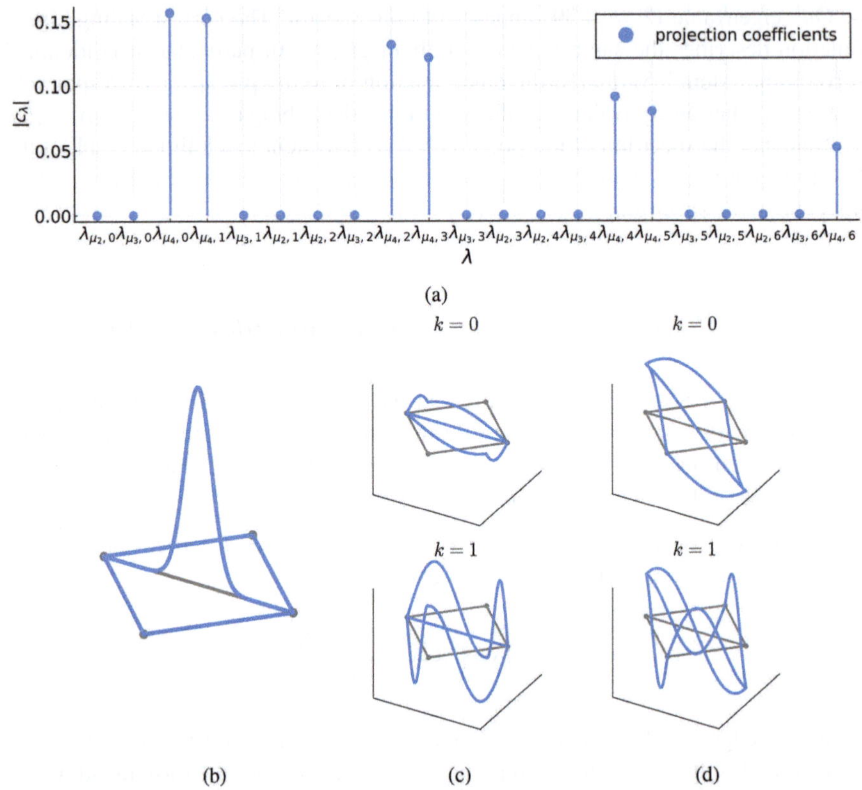

Fig. 6.39 Projection coefficients for Example 6.6 with a selection of associated eigenfunctions. *In subfigure (**a**), the projection coefficients c_λ of the vertex eigenfunctions ϕ_λ corresponding to the eigenvalues $\lambda = \lambda_{\mu,k}$ are plotted for $\mu = \mu_2, \mu_3, \mu_4$ and $k = 0, \ldots, 6$.* (**a**) Projection coefficients of vertex eigenfunctions. (**b**) u on Γ_{dia}. (**c**) ϕ_λ for $\lambda = \lambda_{\mu_2,k}$. (**d**) ϕ_λ for $\lambda = \lambda_{\mu_3,k}$

Chapter 7
Application to the Simulation of Tau Propagation in Alzheimer's Disease

Abstract This chapter discusses the application of the derived methods to the simulation of tau propagation in Alzheimer's disease. The objective is to test the applicability of the methods to real world, large scale data. The main ideas of the *Global Brainsphere Model* and the representation of the brain as a metric graph are introduced. Numerical experiments on the solution of IBVPs governed by linear reaction-diffusion equations are carried out using the finite element discretization from Chap. 3. The eigenvalue algorithm from Sect. 5.3 is applied to compute eigenvalues of metric graphs modeling the brain.

As indicated in the introduction, the motivation for the study of differential equations on metric graphs arose from the interdisciplinary research project *Neurodegeneration Forecasting—A Computational Brainsphere Model for Simulation of Alzheimer's Disease*. The central objective of the project was the development of a *Global Brainsphere Model* (GBM) to predict the propagation of intra-neuronal tau pathology in Alzheimer's disease (AD). In this context, data collected at the Research Center Jülich, Germany, have been utilized to realize a representation of the brain as a metric graph. This representation is a key compartment of the GBM.

The goal of this chapter is to test the applicability of the derived numerical methods to metric graphs that model the brain network. From this, first insights can be gained into the structure of such network types and how a substance spreads across them. Advanced modeling can follow this, but all relevant data and model parameters still need to be collected.

An outline of the GBM has been submitted for a *Snapshot of modern mathematics from Oberwolfach* in joint work with Tolunay Yilmaz and Angela Kunoth [36]. To give some backgrounds on the ideas and motivation at this point, I include parts of this exposition with minor modifications in the first part of Sect. 7.1. Here, the focus is on our approach to modeling the brain network as a metric graph, which I outline with more details than in [36] in Sect. 7.2. Section 7.3 gives a brief description of the data applied to construct a metric graph and initial data. The numerical experiments were carried out for this work and are presented in Sect. 7.4.

A. Weller, *Numerical Methods for Metric Graphs*, Lecture Notes in Mathematics 2382, https://doi.org/10.1007/978-3-032-05011-3_7

7.1 Motivation

Alzheimer's disease (AD) is a slowly progressing neurodegenerative disease causing the brain to shrink and brain cells to die. Much progress is being made in investigating the course and *hallmarks* of the disease: beta-amyloid peptides that aggregate in the surrounding area of the neuronal network (extraneuronal aggregation) and missfolded tau proteins which accumulate in the neurons in form of *tangles* (intraneuronal accumulation) [20, 24]. Tau tangles are believed to affect the functioning of the neurons and eventually to cause their death.

Modern medical imaging techniques nowadays allow to visualize these kinds of plaques in the brain *in-vivo*. Positron emission tomography (PET) scanners examine accumulations of tau and beta-amyloid and, additionally, functional magnetic resonance imaging (fMRI) and diffusion tensor imaging (DTI) provide indications on the structure and functioning of a human brain.

With the help of these techniques, a lot of studies have been made on the behavior of tau and beta-amyloid as well as connectivity patterns in the brain. The prevailing hypothesis investigated in connection with tau is the *transneuronal* spread hypothesis [19]. By transneuronal, we mean that tau tangles travel from neuron to neuron, inducing tangles in neighboring neurons in a *prion* (protein infection) like fashion. This hypothesis motivates the simulation of tau propagation as diffusion process on the brain network. But, on the other hand, tau tangles are also believed to disrupt the connections between neurons and thereby to influence the brain network architecture.

The interplay between tau and network connectivity has become an active field of research with the modeling of the brain as a (combinatorial) graph as one of the methods in focus. In 2018, Cope et al. published a novel approach to investigate tau patterns and their dependency on network connectivity measures, such as weighted degree of the vertices [16]. This study has been reproduced by our interdisciplinary research team [55] and similar results have been obtained: high functional connectivity of a network node correlates with high measured tau pathology in AD.

On the other hand, a subsequent study of our group focused on the effects of tau on brain connectivity, more precisely, on community formation in the brain [56]. Our findings suggest that not only the brain network contributes to the tau distribution but also that neurodegeneration affects the connectivity patterns of the brain. This implies that a diffusion model of transneuronal spread should be capable of dealing with altered graphs.

As already mentioned in the introduction, Raj et al. study a *network heat equation* on a combinatorial graph to investigate pathology patterns in the brain of AD patients. In contrast to the problems considered during this work, the diffusion is modeled on a discrete domain (in contrast to a metric graph). The network heat equation can then be solved explicitly by an eigendecomposition of the graph Laplacian matrix. Therefore, in their study, Raj et al. identify the first eigenvectors (i.e., the eigenvectors belonging to the smallest eigenvalues) of the graph Laplacian matrix as main drivers of diffusion and compare them with measured atrophy.

However, the model is not applied to actually model propagation on the brain network. Moreover, it is a pure diffusive model and no additional phenomena such as production or aggregation of tau proteins are included. There are some other attempts that extend the pure diffusive models, for example [47, 51], but to my knowledge, no continuous model (i.e., applying metric graphs) was investigated so far.

In a nutshell, the objective of the GBM is to analyze and combine the various *in-vivo* data in a mathematical model to simulate and predict future course of AD. In addition to the interaction of tau and the brain network, the presence of beta-amyloid plaques in the surrounding of a neuron could influence its tendency to aggregate tau. The GBM should thus comprise a dynamical brain network model, a tau propagation and a beta-amyloid distribution model. We will here concentrate on the simulation of tau propagation based on a diffusion-type distribution on the brain network. As the latter serves as a basis for this model, we will present our current approach of a network (or graph) representation of the brain in the following section.

7.2 Brain Network Model

Our aim is to represent the human brain as network consisting of vertices and edges. Unfortunately, the resolution of common MRI scanners is not high enough to render the extremely small neurons and synapses in the brain. The most prevalent method to obtain a network representation comprises thousand of neurons to one brain region of interest (ROI). If we consider the brain as three-dimensional object, the definition of these ROIs corresponds to a partitioning into a number of contiguous regions.

In our case, this resulted in the parcellation displayed in Fig. 7.1 with 571 regions with an approximate volume of 2 ml (the *UoC-atlas* has been computed by Philipp Schlüter as outlined in [56]). In particular, the ROIs in our atlas all have approximately the same size and shape. Each of these regions then serves as a vertex in our graph, where we position the vertex in the middle of the brain region, see Fig. 7.1.

Remark 7.1 In fact, the parcellation in 571 regions is already very fine. As a comparison, Raj et al. have used 116 regions in their network diffusion study [45]. Moreover, their brain regions vary in size and shape. The idea to use regions with an approximate volume of 2 ml originated from [16].

Having defined the brain regions, it remains to specify how these regions are connected to each other. As the application of graph theory in neuroscience has become prominent, there exist several established methods to define connectivity. Among the two most prevalent ones are structural and functional connectivity.

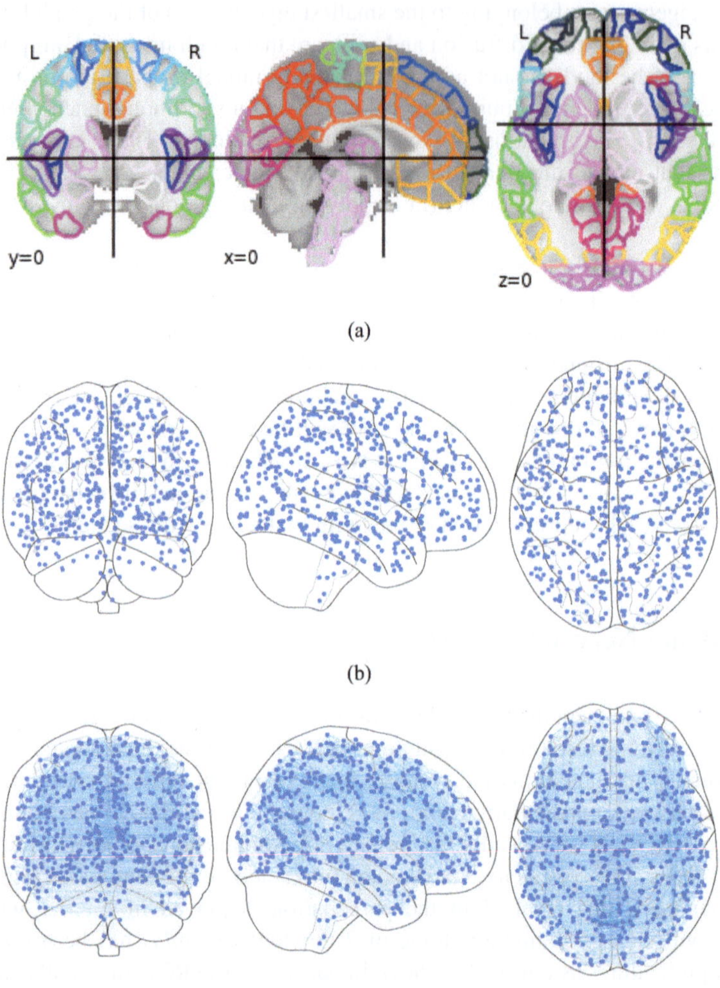

(a)

(b)

(c)

Fig. 7.1 Construction of the brain network model. *Graphical representation using* `nilearn` *in python*. (**a**) UoC brain atlas with 571 brain regions. (**b**) Vertices of the brain network. (**c**) Vertices and edges of the brain network

Structural Connectivity. The relative diffusion coefficient of water molecules in the brain can be measured in each voxel using diffusion tensor imaging (DTI). This coefficient provides information about the existence and strength of a physical connection between two brain regions [40].

Functional Connectivity. The concept of functional connectivity aims to mirror the functional relationship between two brain regions, or in other words, their extend of interaction. During a resting-state functional magnetic resonance imaging (rs-fMRI) session, patients are instructed not to move or think about anything in particular while a series of scans are taken. To analyze the MRI, one makes use of the effect that the signal of an MRI depends on the blood oxygen level of a certain region in the brain. The blood oxygen level in turn reflects the activation level of that region [14]. Recording a series of images of a subject over time delivers blood oxygen level dependent (BOLD) time-series for each brain region whose pairwise correlation then determine the (undirected) functional connectivity matrix.

Functional connectivity matrices have been studied in detail by our group and me for the study [56]. Structural connectivity was studied intensively and computed by a bachelor student.[1] This study applied the UoC-atlas but some difficulties arose due to the small size of the atlas regions and their proper alignment with a standard space. The data therefore need to be revised before they can be applied in the diffusion model.

7.3 Data Description

7.3.1 Metric Graph Model of the Brain Network

In [56], we studied functional connectivity matrices for AD patients as well as for a young and age matched healthy control cohort. In total, data of twenty AD patients and a group of twenty age-matched healthy controls who underwent an MRI scanning protocol at the Research Center Jülich, Germany, were analyzed. In addition, fMRI data from a group of 24 young healthy controls from a public available rs-fMRI dataset by Berlin-Margulies [46] were included. Details on data acquisition and processing can be found in the “Methods” section in [56]. The functional connectivity matrix is typically thresholded such that the strongest 5–10% connections determine the adjacency matrix of the graph. For the following experiments, we applied a local thresholding method with 6% network density as described in [56].[2]

A group-average brain network adjacency matrix is computed for each group (AD, control and age-matched control) as the average of the thresholded functional connectivity matrices, see Fig. 7.2. To obtain a metric graph from the adjacency matrices, we equip the edges with a length according to the euclidian distance of its start and end vertex (Fig. 7.3). We will denote the constructed metric graphs by Γ_{AD}, Γ_{Control}, and $\Gamma_{\text{Age-matched}}$.

[1] K. Kolb: Analyse der strukturellen Konnektivität im Gehirn im Rahmen des Projekts Neurodegeneration Forecast: A Computational Brainsphere Model for Simulation of Alzheimer’s Disease, Bachelor thesis, University of Cologne (2020).

[2] Thresholding was conducted with the `Maybrain` package in `python` (github.com/Rittman/maybrain).

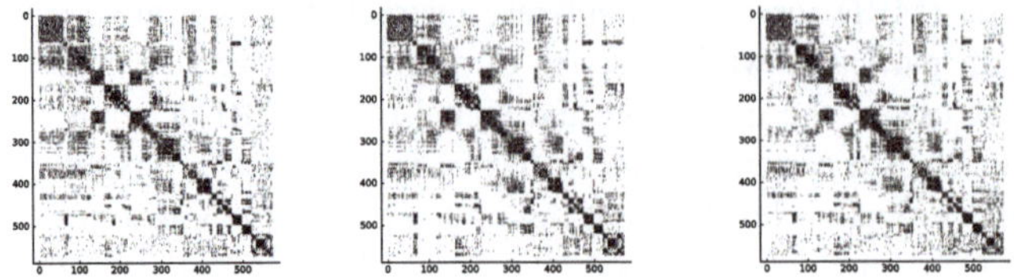

(a) (b) (c)

Fig. 7.2 Adjacency matrices of $\Gamma_{\text{Age-matched}}$, Γ_{Control}, and Γ_{AD}. *Black markers indicate an edge in the graph.* (**a**) Control. (**b**) Age-matched control. (**c**) Alzheimer's disease

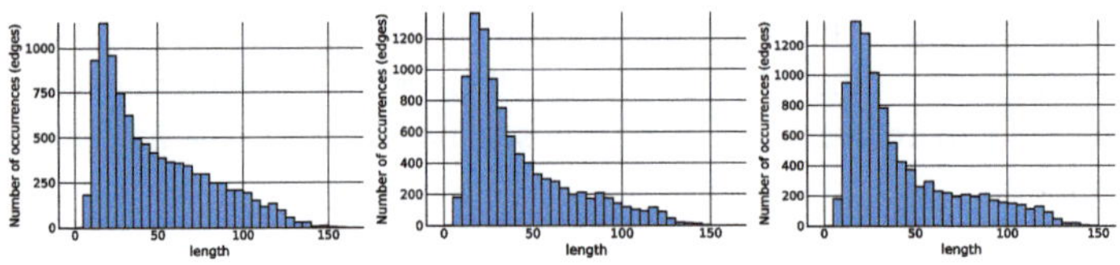

(a) (b) (c)

Fig. 7.3 Distribution of the edge lengths in $\Gamma_{\text{Age-matched}}$, Γ_{Control}, and Γ_{AD}. (**a**) Control. (**b**) Age-matched control. (**c**) Alzheimer's disease

All data applied for the construction of the metric graph have been originally studied and preprocessed for [56], and a detailed description of the applied methods can be found therein. The original data are stored in the Multimodal Imaging Cologne group according to their ethics approved data protection policies.

7.3.2 *Tau Initial Data*

A particular type of tracer for positron emission tomography (PET) can be used to assess the distribution of tau pathology in vivo. For a collaboration with the Multimodal Neuroimaging Cologne group, tau pathology data from ADNI (Alzheimer's Disease Neuroimaging Initiative)[3] have been analyzed [31]. In particular, longitudinal PET data are available for the studied patient group, meaning that the patients underwent a tau PET scan at two different time points, monitoring the deviation.

In total, data of 98 amyloid-positive patients for which both a baseline amyloid (18F-AV45) scan and a longitudinal tau (18F-AV1451) PET scan were available have been selected. In addition, 35 amyloid-negative patients were used as reference

[3] ADNI is a public-private partnership with the primary goal to test the combination of multimodal imaging methods to measure the progression of mild cognitive impairment (MCI) and early AD, see www.adni-info.org.

to compute z-transformed tau and annual tau change maps (compare [30] or [8] for details on the data preprocessing). The z-values are then extracted for each of the 571 brain regions of the *UoC-atlas*.

The 98 amyloid-positive patients are divided into 3 groups: 48 amyloid-positive cognitively normal subjects (group 1), 35 subjects with mild cognitive impairment (group 2) and a group of 15 AD patients (group 3). For each group, an average baseline tau pattern is then computed as the average of the positive z-values across the patients, compare Fig. 7.4.

Additionally, a deviation pattern is computed on the basis of the annual tau change maps, see Fig. 7.5. The original data applied for the generation of initial data originate from the ADNI database. The data were preprocessed by the Multimodal Imaging Cologne group using in-house scripts that are stored according to their policies.

7.4 Numerical Experiments

The previously described data are now used to apply the derived numerical methods to large-scale, real world data. In the first subsection, reaction-diffusion equations on the brain network of the healthy control cohort, i.e., Γ_{Control} will be solved with the finite element method. In the second subsection, a lower part of the spectrum of Γ_{Control}, $\Gamma_{\text{Age-matched}}$, and Γ_{AD} is computed using equilateral approximations followed by a Newton-trace iteration.

The numerical experiments constitute a first attempt of investigation diffusion-type models on brain networks. For sophisticated simulations, additional data have to be assessed and all relevant model parameters have to be identified and determined in collaboration with the colleagues from the Department of Nuclear Medicine.

7.4.1 Finite Element Approximation of Reaction-Diffusion Equations on the Functional Brain Network

The baseline tau patterns described in Sect. 7.3.2 serve as initial condition on the network nodes in the diffusion model, i.e., we consider three different initial conditions $u_{\mathcal{V}}^{0,G1}$, $u_{\mathcal{V}}^{0,G2}$ and $u_{\mathcal{V}}^{0,G3}$, each representing the initial tau pattern of one patient group (as illustrated in Fig. 7.4). A particular difficulty in the modeling is the definition of a suitable initial condition on the whole metric graph. As described in the previous section, initial tau pathology can only be measured in the vertices (i.e., ROIs) themselves and not along the edges. The data therefore have to be interpolated

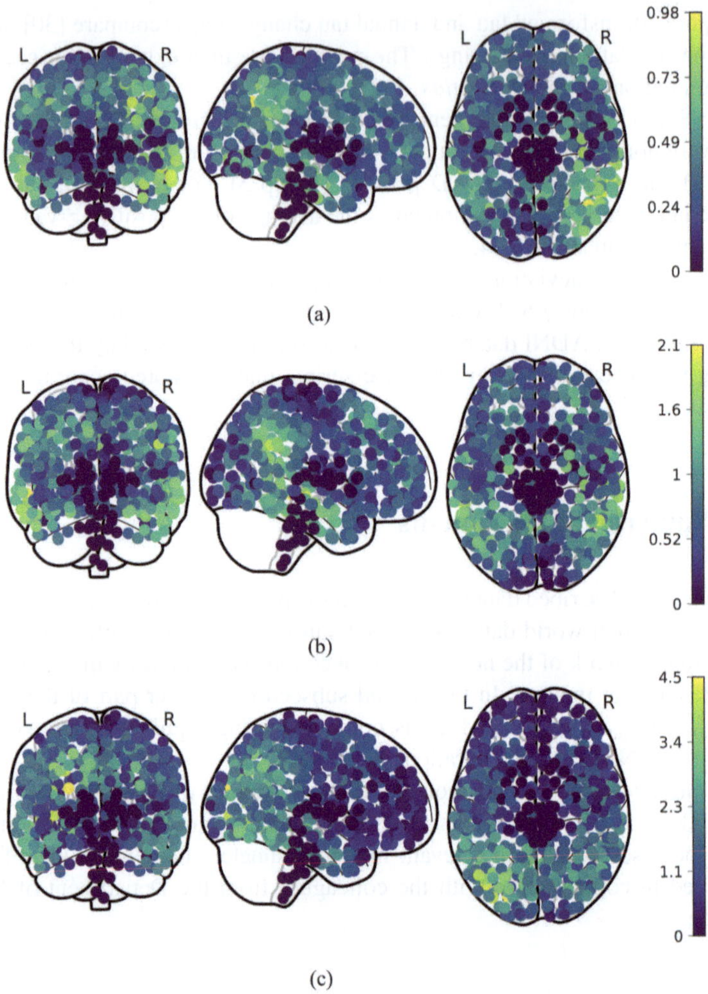

Fig. 7.4 Initial tau patterns measured in baseline PET scan. *Graphical representation using* `nilearn` *in python.* Note that a z-value > 1.65 already indicates a significant uptake of tau pathology. We therefore additionally provide thresholded images of the initial tau patterns in the appendix (Fig. 7.11). (**a**) Group 1 (amyloid-positive cognitively normal). (**b**) Group 2 (mild cognitive impairment). (**c**) Group 3 (AD)

at the interior of the edges. For given data $u^0_{\mathcal{V}} \in \mathbb{R}^n$ on the vertices $\mathcal{V}$, we propose to define

$$u^0_e(x) = (u^0_{\mathcal{V}}(t(e)) - u^0_{\mathcal{V}}(o(e)))\frac{\exp(-\ell_e/x)}{\exp(-\ell_e/x) + \exp(-\ell_e/(\ell_e - x))} + u^0_{\mathcal{V}}(o(e)) \tag{7.1}$$

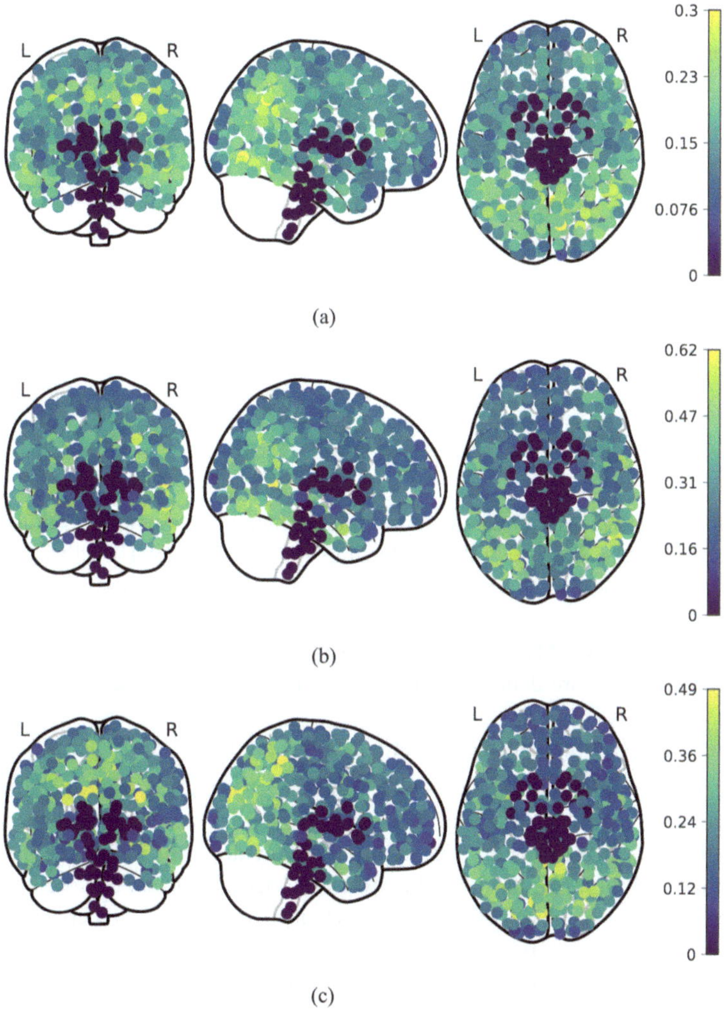

Fig. 7.5 Tau deviation as measured in follow-up PET scan. *Graphical representation using* `nilearn` *in python*. (**a**) Group 1 (amyloid-positive cognitively normal). (**b**) Group 2 (mild cognitive impairment). (**c**) Group 3 (AD)

for $x \in (0, \ell_e)$ where $o(e)$ and $t(e)$ are the start respectively end vertex of edge e.

To streamline the exposition, we outline the results for group 2, i.e., we set $u^0 = u^{0,\mathrm{G2}}$. Some results for the other groups are provided in the appendix (Figs. 7.12 and 7.13).

We start with the simplest model, a pure diffusion model. To be more precise, the propagation of tau pathology on Γ_{Control} is modeled by the IBVP

$$\frac{\partial u}{\partial t} + \mathcal{H}u = 0 \quad \text{on} \quad \Gamma \times [0, T] \tag{7.2}$$

under Neumann-Kirchhoff conditions and with initial condition $u^0 = u^{0,\text{G2}}$, compare Problem 2.16. In addition, we consider a weighted version of (7.2) that accounts for different connection strengths of the edges. In this case, the differential operator $\mathcal{H}_\beta$ acts on u_e as $\mathcal{H}_\beta : u_e \mapsto -\beta_e \frac{\partial^2}{\partial x^2} u_e$ for $e \in \mathcal{E}$. The edge specific diffusion coefficients $\beta_e \in \mathbb{R}^+$ are chosen as the functional connectivity strength. The weighted diffusion equation then reads

$$\frac{\partial u}{\partial t} + \mathcal{H}_\beta u = 0 \quad \text{on} \quad \Gamma \times [0, T]. \tag{7.3}$$

The solution of (7.2) and (7.3) are approximated by a finite element semidiscretization with step size $h_{\max} = 0.0625$ (leading to 10504151 DoF) followed by the Crank-Nicolson method (3.15) with step size $\Delta t = h$. The arising SLEs are solved with a matrix-free implementation of the multigrid algorithm (Algorithm 2) derived in Sect. 3.3.2 (CN-MGM). The values of the finite element solution of (7.2) at the vertices $\mathcal{V}$ at time t are denoted by $\mathbf{u}_{\mathcal{V}}^t$ and by $\mathbf{u}_{\mathcal{V},\beta}^t$ for the finite element solution of (7.3).

In Fig. 7.6, we illustrate the deviation of the predicted tau patterns $\mathbf{u}_{\mathcal{V}}^t$, $\mathbf{u}_{\mathcal{V},\beta}^t$ and the initial condition, i.e., $\mathbf{u}_{\mathcal{V}}^t - \mathbf{u}_{\mathcal{V}}^{0,\text{G2}}$ and $\mathbf{u}_{\mathcal{V},\beta}^t - \mathbf{u}_{\mathcal{V}}^{0,\text{G2}}$ at time $t = 12.5$. The differences are marginal, however, we can observe that some regions are effected less by the weighted diffusion, suggesting that they are only weakly connected to the regions with high tau burden. The diffusion coefficients can thus be applied as parameters to control the diffusion to regions that are less likely to receive pathology.

The previous results obtained by a pure diffusion equation cannot model the overall tau uptake measured in the follow-up PET scan (compare Fig. 7.5). For $\alpha \in \mathbb{R}$, we therefore next consider the linear reaction-diffusion equation

$$\frac{\partial u}{\partial t} + \mathcal{H}u = \alpha u \quad \text{on} \quad \Gamma \times [0, T]$$

under Neumann-Kirchhoff conditions and with initial condition $u^0 = u^{0,\text{G2}}$, compare Problem 2.17. The approximate solution at time t will be denoted by $\mathbf{u}_{\mathcal{V},\alpha}^t$ and is computed with the same pipeline as for the diffusion equation, i.e., a finite element approximation ($h = 0.0625$) followed by the CN-MGM solver ($\Delta t = h$). We illustrate the actual observed deviation (follow-up PET scan) from the initial data together with the deviation of the predicted tau patterns $\mathbf{u}_{\mathcal{V},\alpha}^t$ for $\alpha \in \{0.001, 0.005\}$ in Fig. 7.7.

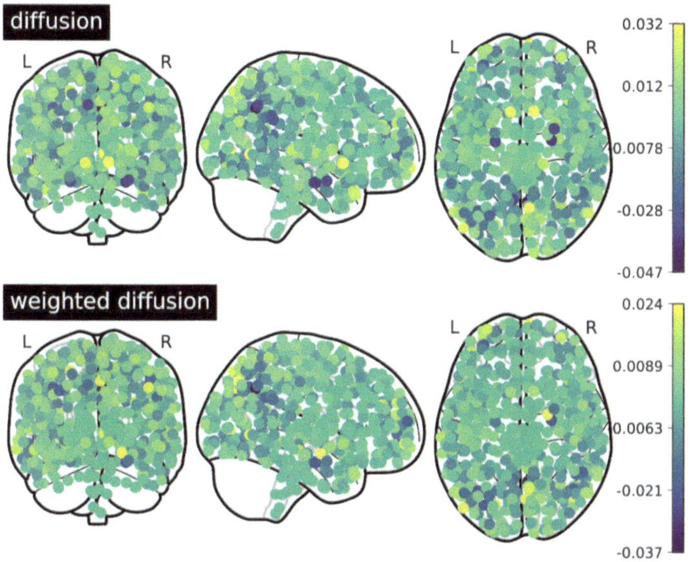

Fig. 7.6 Deviation $\mathbf{u}^t_{\mathcal{V}} - \mathbf{u}^{0,\mathrm{G2}}_{\mathcal{V}}$ and $\mathbf{u}^t_{\mathcal{V},\beta} - \mathbf{u}^{0,\mathrm{G2}}_{\mathcal{V}}$ of the predicted tau pattern and the initial data at $t = 12.5$. *Graphical representation using* `nilearn` *in python*

For each α, we have chosen a time point t where the overall amount of deviation matches the observed deviation. Clearly, this state is attained faster for larger reaction terms.

The illustration of the results suggests that the tau deviation might be better explained by more reaction-dominated equations. However, recall that we are considering annual tau change maps, long-term effects might be subject to more diffusion or additional effects. Furthermore, we point out that structural connectivity networks might be better suited for the modeling of a diffusion process. The computation of these will therefore be an important next step. Moreover, in the very simplified model above, we have defined a global reaction term acting on each edge in the same manner. Local reaction terms identified from collected data and properties of the different brain regions should be derived for future experiments and can help to better capture the observed behavior.

7.4.2 *Spectra of Functional Connectivity Graphs*

So far, we have only considered the metric graph Γ_{Control} obtained from a young healthy control cohort. However, as indicated earlier, we expect the connectivity patterns to be altered with age and disease progression. To support this assumption, we will in this subsection compare the spectra of the brain metric graphs obtained from the three different groups, i.e., $\Gamma_{\text{Control}}, \Gamma_{\text{Age-matched}}, \Gamma_{\text{AD}}$.

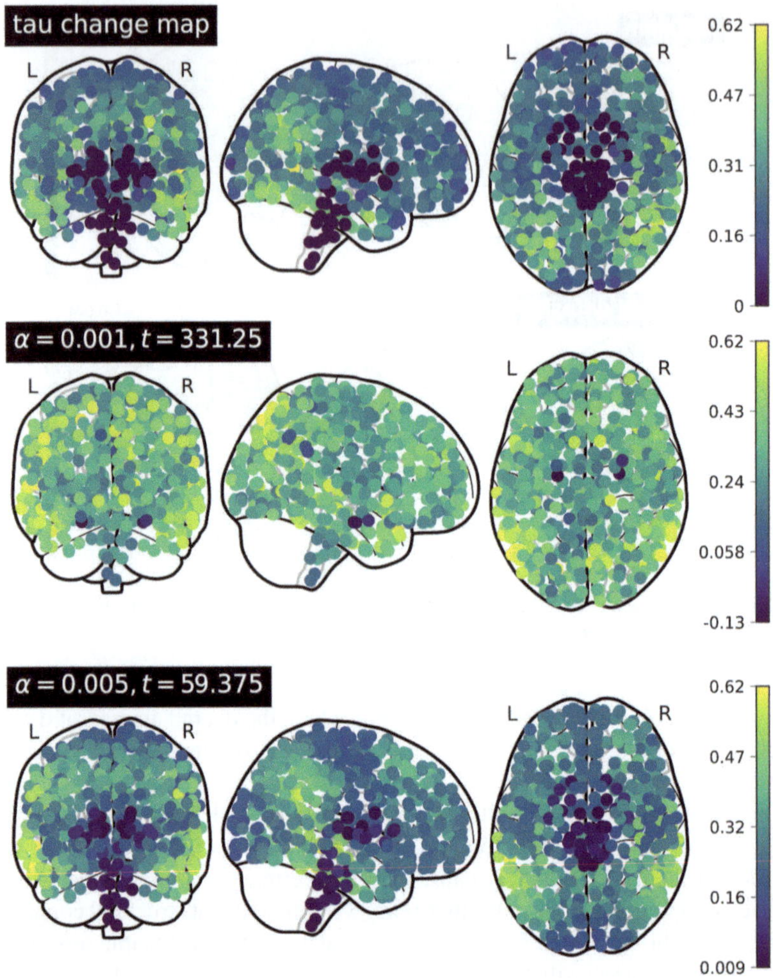

Fig. 7.7 Actual tau change map and deviation $\mathbf{u}^t_{\mathcal{V},\alpha} - \mathbf{u}^{0,\mathrm{G2}}_{\mathcal{V}}$ of the predicted tau pattern and the initial data. *Graphical representation using* `nilearn` *in python*

To compute the quantum graph spectra, we first round the edge lengths to two decimal digits and then compute the first 20 eigenvalues of equilateral floor and ceil approximations $\mathfrak{G}_{\mathrm{fl},h}$, $\mathfrak{G}_{\mathrm{cl},h}$ with the nested iteration algorithm (Algorithm 7). As the lowest discretization level, we choose an equilateral approximation with edge length $h = 2$, the finest level corresponds to an approximation with edge length $h = 0.5$. Thus, we intend to employ the eigenvalues of the equilateral approximation $\lambda_q(\mathfrak{G}_{\mathrm{fl},0.5})$ and $\lambda_q(\mathfrak{G}_{\mathrm{cl},0.5})$ as estimates for the initial guess of the Newton-trace iteration.

First, we verify the estimates $\lambda_q(\mathfrak{G}_{\mathrm{cl},0.5})$ and $\lambda_q(\mathfrak{G}_{\mathrm{fl},0.5})$ for the first six (non-zero) eigenvalues by plotting the reverse condition number, compare Fig. 7.8. As can

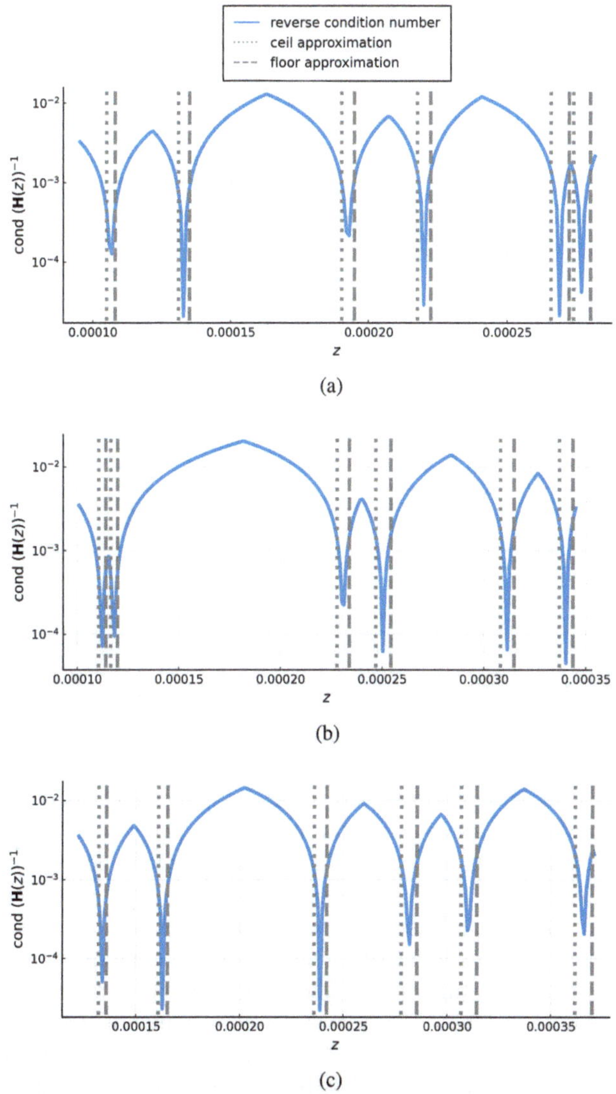

Fig. 7.8 First five eigenvalues of $\mathfrak{G}_{\mathrm{cl},0.5}$ and $\mathfrak{G}_{\mathrm{fl},0.5}$ as upper and lower bounds for the roots of the reverse condition number $\mathrm{cond}(\mathbf{H}(z))^{-1}$. **(a)** Control. **(b)** Age-matched control. **(c)** Alzheimer's disease

be seen, the roots of the reverse condition number lie exactly between the estimates obtained from the equilateral floor and ceil approximations, and, in fact, the Newton-trace iteration converges to the desired eigenvalues, compare Table 7.1.

A comparison of the computed eigenvalues for the three different graphs reveals differences between the three groups, see Fig. 7.9. Moreover, we have plotted the

Table 7.1 Eigenvalues of Γ_{Control}, $\Gamma_{\text{Age-matched}}$, and Γ_{AD} computed by Newton-trace

λ_q	Control	Age-matched control	AD
$q = 2$	1.07e−04	1.12e−04	1.34e−04
$q = 3$	1.33e−04	1.18e−04	1.63e−04
$q = 4$	1.93e−04	2.31e−04	2.39e−04
$q = 5$	2.20e−04	2.51e−04	2.82e−04
$q = 6$	2.69e−04	3.11e−04	3.10e−04
$q = 7$	2.77e−04	3.40e−04	3.65e−04

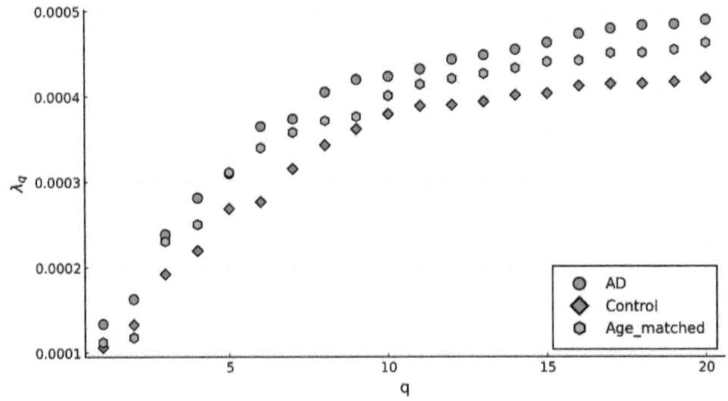

Fig. 7.9 First 20 (non-zero) eigenvalues of Γ_{AD}, Γ_{Control}, and $\Gamma_{\text{Age-matched}}$

values of the first two eigenfunctions (corresponding to non-zero eigenvalues) ϕ_2, ϕ_3 on the vertices for each group in Fig. 7.10. However, the interpretation of the results is not straightforward. In general, smaller eigenvalues indicate that the diffusion progresses more slowly throughout the network. But, as observed in the numerical experiments in Sect. 6.3.3, this also critically depends on the initial condition. When it comes to the values of the eigenfunctions at the vertices, it will be interesting to investigate if they can be interpreted as an analog to the Fiedler's vector in combinatorial graph theory (to follow up on the interpretation in [45]).

Overall, the observed differences indicate that the network architecture of the age-matched control and the AD cohort is altered in comparison to young controls, and, that these alterations influence the distribution of pathology on the brain network. The results indicate that alterations should be integrated in the model, for instance by a dynamical model where the underlying graph is subject to age and disease related changes. At this point, we would also like to point out that the applied thresholding scheme critically influences the network structure. Since we applied a relative thresholding in each group, it remains to investigate how the results are comparable between the groups. Different thresholds and thresholding strategies will be applied to analyze if the observed changes are robust towards thresholding.

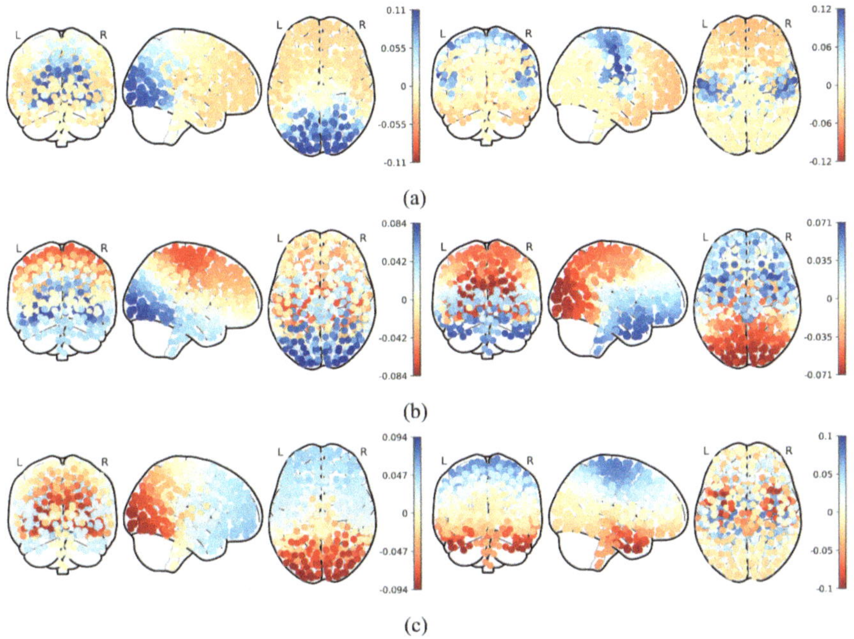

Fig. 7.10 Values of ϕ_2 and ϕ_3 at the vertices of Γ_{Control}, $\Gamma_{\text{Age-matched}}$, and Γ_{AD}. *Graphical representation using* `nilearn` *in python*. (**a**) Control. (**b**) Age-matched control. (**c**) Alzheimer's disease

Appendix

Simulation of Tau Propagation

See Figs. 7.11, 7.12, and 7.13.

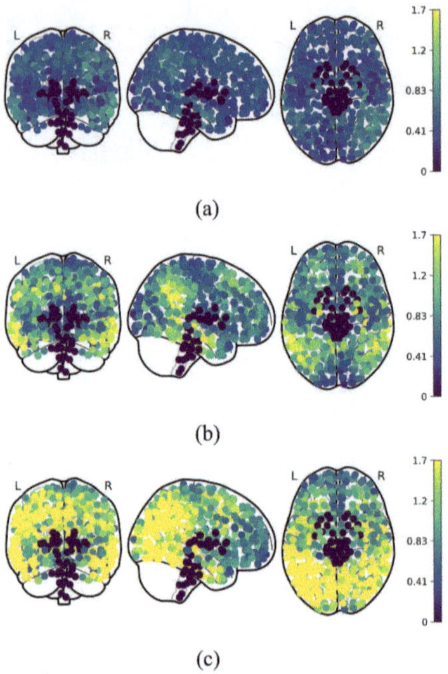

Fig. 7.11 Initial tau patterns measured in baseline PET scan with z-value threshold 1.65. *Graphical representation using* `nilearn` *in python*. (**a**) Group 1 (amyloid-positive cognitively normal). (**b**) Group 2 (mild cognitive impairment). (**c**) Group 3 (AD)

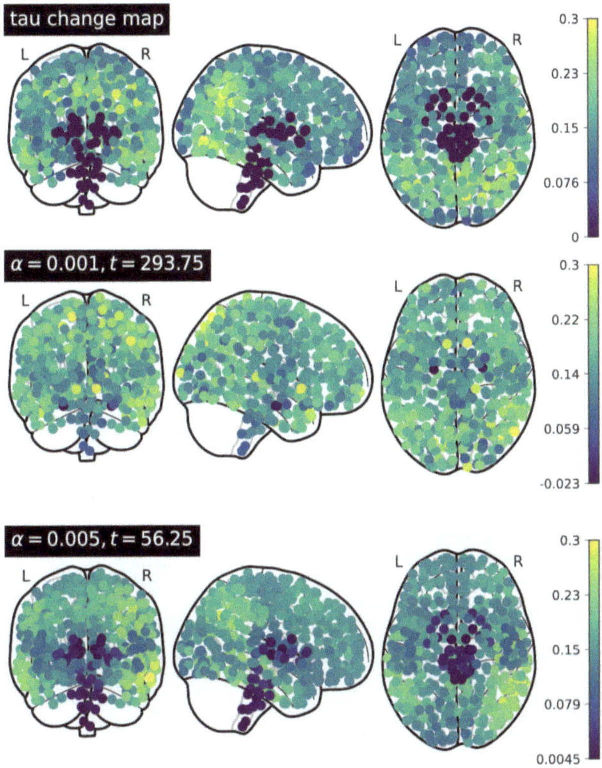

Fig. 7.12 Group 1: Actual tau change map and deviation $\mathbf{u}^t_{\mathcal{V},\alpha} - \mathbf{u}^{0,\mathrm{G1}}_{\mathcal{V}}$ of the predicted tau pattern and the initial data. *Graphical representation using* `nilearn` *in python*

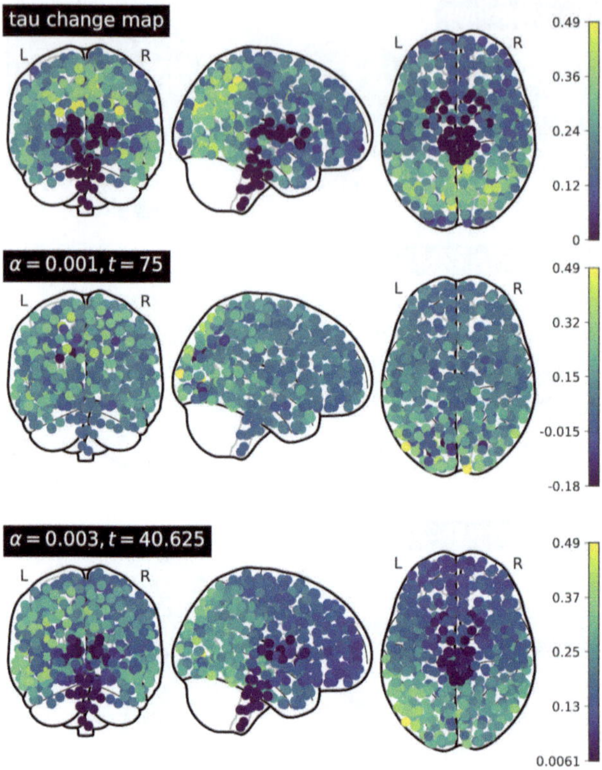

Fig. 7.13 Group 3: Actual tau change map and deviation $\mathbf{u}^t_{\mathcal{V},\alpha} - \mathbf{u}^{0,\mathrm{G3}}_{\mathcal{V}}$ of the predicted tau pattern and the initial data. *Graphical representation using* `nilearn` *in python*

Chapter 8
Conclusion

Abstract The main concepts studied in this book are summarized. This is followed by a brief discussion of how to use the extended graph framework in the context of finite difference methods for metric graphs. Some aspects of the spectral method are again critically reviewed and impulses are given for potential future work.

8.1 Summary

Let us briefly summarize the expositions of the previous chapters. First, we derived a formulation of PDEs posed on metric graphs by introducing Neumann-Kirchhoff coupling conditions. We focused on parabolic differential equations evolving the negative second spatial order derivative. The foundation for both of the proposed solution methods is a weak formulation of the PDE of interest which was established for a linear parabolic problem as well as for a semilinear reaction-diffusion equation. The well-posedness of the weak formulations was deduced by standard requirements on the bilinear form. Moreover, an important finding in the first part of this book was the self-adjointness of the differential operator $\mathcal{H}$. This property is essential for the practicability of the spectral solution method considered later since it assures the eigenfunctions of $\mathcal{H}$ to span an orthogonal basis of the solution space.

As a first solution approach, we reviewed the finite element approximation presented in [2]. We put some effort into the detailed discussion of the concept of extended graphs and the structure of their graph Laplacian and incidence matrix. This has been essential since it later allowed a convenient representation of the finite element stiffness and mass matrix which is not only practical for representing the matrices, but also allows an efficient assembling using the incidence matrix of the extended graph. The latter, in turn, can be constructed from the incidence matrix of the original graph using Kronecker products.

The obtained quadratic convergence rate of the finite element discretization is supported by the conducted numerical experiments. We proposed implicit-explicit time stepping methods to solve the system of ordinary differential equations arising from the semidiscretization. Here, the implicit part of the scheme gives rise to a system of linear equations. For the solution of these systems, we introduced

A. Weller, *Numerical Methods for Metric Graphs*, Lecture Notes in Mathematics 2382, https://doi.org/10.1007/978-3-032-05011-3_8

a multigrid iteration and also outlined a domain decomposition approach in the appendix.

Moreover, we derived a spectral Galerkin method with a trial space composed of eigenfunctions of the differential operator acting on $\mathrm{dom}_{\mathcal{H},\mathrm{NK}}$. In other words, the objective was to express the solution as a truncated eigenfunction expansion. We have seen that spectral accuracy can be obtained for a class of functions that fulfill the vertex coupling conditions together with each of its even derivatives. This means that the truncation error decays faster than any polynomial with the addition of further basis functions. These findings were as well supported by numerical experiments.

Given the particular structure of the mass and stiffness matrix, the solution of the discretized system can be computed directly by using the matrix exponential, or by exponential integrators in the semilinear case. Rather, the main costs of the spectral Galerkin method arise from the computation of an eigenfunction basis as well as the evaluation of inner products on graphs. The latter is a specific difficulty in spectral methods given that the basis functions have global support. Thus, integrals over the whole metric graph have to be evaluated, for instance to compute orthogonal projections. We addressed this with a Filon-type quadrature formula which takes advantage of the fact that we have a closed form representation of the eigenfunctions.

For the computation of an eigenfunction basis, we proposed a method relying on a reduction of the continuous problem to a (non-)linear eigenvalue problem on the underlying combinatorial graph. In this context, equilateral graphs were of special interest since the NEP in this case further simplifies to a linear eigenvalue problem. This remarkable observation allowed to directly relate the eigenvalues and eigenfunctions of a metric graph to the eigenvalues and eigenvectors of its underlying combinatorial graph. However, the relation does not hold for some special *non-vertex eigenvalues* such that some more work had to be done to resolve the corresponding eigenfunctions. For this purpose, we proposed a novel approach relying on the insertion of artificial vertices on the edges.

Having covered the equilateral case, we moved on to general, non-equilateral metric graphs. By a standard approach, the NEP was reformulated to a root finding problem and solved by a Newton-trace iteration. The novelty of our proposed method is the application of the Newton-trace iteration to initial estimates obtained from *equilateral approximations* of the non-equilateral graph, ensuring fast convergence.

Finally, two test problems were considered to compare the finite element and spectral Galerkin method in terms of convergence and complexity. Moreover, the applicability of the methods to real world data obtained during the research project on Alzheimer's disease was tested. The first results suggest that reaction terms play an important role in the modeling of disease propagation.

8.2 Discussion and Future Work

One of the main purposes of this work was the detailed derivation and presentation of numerical methods for parabolic PDEs on metric graphs. To streamline the exposition, I have decided to not include a finite difference method in order to work exclusively with weak formulations. This guaranteed better comparability of the presented methods by an error measured in the $L^2(\Gamma)$-norm. However, the finite difference method introduced in [10] can compete with the finite element method in practice and also exhibits a quadratic error reduction (measured in the $\|\cdot\|_\infty$-norm).

In order to increase the practicality of the finite difference method, I have, in particular, further extended the approach by the representation of the discretization matrices in terms of the extended graph, just as in the finite element approach. The finite difference discretization of the generalized heat equation then leads to the system of ODEs

$$\frac{\mathrm{d}}{\mathrm{d}t}\mathbf{u}(t) + 2\hat{\mathbf{Q}}^{-1}\hat{\mathbf{L}}\mathbf{u}(t) = \hat{\mathbf{f}} \tag{8.1}$$

where $\hat{\mathbf{Q}} := \operatorname{diag}\left(\{(|\tilde{\mathbf{N}}\tilde{\mathbf{W}}^{-1}\tilde{\mathbf{N}}^T|)_{i,i}\}_{i=1}^{\tilde{n}}\right)$ and $\tilde{\mathbf{N}}$, $\hat{\mathbf{L}}$ are the incidence and weighted graph Laplacian matrix of the extended graph as in Definition 3.5. Interestingly, for equilateral graphs, the discretization matrix $2\hat{\mathbf{Q}}^{-1}\hat{\mathbf{L}}$ simplifies to $\Delta_{\mathcal{G}}$ which is our well known harmonic graph Laplacian matrix that played an important role in the determination of quantum graph spectra. The system (8.1) has a very similar structure to the finite element semidiscretization with the advantage that the equivalent to the mass matrix is the identity. For the solution of (8.1), it is therefore of interest to work with an approximation of the matrix exponential $\exp(-t\,2\hat{\mathbf{Q}}^{-1}\hat{\mathbf{L}})$ (for example occurring in exponential integrator methods) instead of a time stepping method. Note that in the finite element setting, the same idea gives rise to the computation of $\exp(-t\hat{\mathbf{M}}^{-1}\hat{\mathbf{L}})$ where $\hat{\mathbf{M}}$ (in contrast to $\hat{\mathbf{Q}}$) is not diagonal. This can, however, be avoided if the mass matrix approximation (Lemma 3.7) is exploited.

When it comes to the spectral Galerkin method, the main computational costs are caused by the computation of an eigenfunction basis. In the equilateral case, the algorithm derived in Chap. 5 reduces this problem to the solutions of discrete eigenvalue problems. The efficient computation of eigenvalues and eigenvectors of graph Laplacian matrices is therefore of great interest. In particular, the systems required for the computation of the non-vertex spectrum arise from an extended graph Laplacian matrix. For these, the development of a multigrid method is worth considering. This can then also be used to accelerate the nested iteration approach to compute initial solutions for the Newton-trace iteration in the non-equilateral case. In the context of non-equilateral spectral computations, our findings suggest the possibility of approaching the eigenvalues up to arbitrary precision by using an equilateral floor or ceil approximation of the exact graph. It remains to further investigate this assumption from a theoretical point of view. The development of corresponding error estimates should be addressed in future work.

As a follow-up to the initial results of the spectral analysis, we point out that a further study of the behavior of eigenfunctions will be interesting with regard to approximation theory on metric graphs. In particular, this will help to determine eigenfunctions that are important to resolve a given function on a graph a priori (i.e., before computing the projection coefficients).

A further improvement of the computational complexity can be achieved through the revision of the Filon quadrature rule for the computation of inner products. Polynomials of higher order can be used and, in the non-equilateral case, a reasonable exploitation of function evaluations by a careful choice of the quadrature points has to be implemented.

Within the comparison of the spectral and finite element method, we want to point out that in the more general case of the generalized heat and reaction-diffusion equations, additional computational costs arise in both methods due to the computation of the inner products on the right-hand side. Moreover, in the non-linear case, a direct solution of the semidiscretized system is no longer possible for the spectral method and exponential integrators have to be applied instead. A comparison of the computational costs for different parameters (for example right-hand side, time T, non-linearity, initial condition) are interesting subjects for future work.

Finally, we would like to note out that one of the disadvantages of the spectral method is that it was specifically designed for $\mathcal{H}$ acting on $\mathrm{dom}_{\mathcal{H},\mathrm{NK}}$, i.e., for Neumann-Kirchhoff coupling conditions. The latter have guaranteed the self-adjointness of $\mathcal{H}$. In fact, there are other possible coupling conditions that entail a self-adjoint operator, see for example [6], Theorem 1.4.4. When considering different coupling conditions, a new method for the computation of spectral basis functions has to be developed. However, Neumann-Kirchhoff coupling conditions are certainly the most prominent coupling conditions and the discussed spectral approach naturally covers several PDEs, including fractional diffusion equations.

References

1. Ainsworth, M., Weller, A.: A spectral Galerkin method for the solution of partial differential equations on metric graphs. Oberwolfach Reports, Workshop Report 36 (2021)
2. Arioli, M., Benzi, M.: A finite element method for quantum graphs. IMA J. Numer. Anal. **38**(3), 1119–1163 (2018). https://doi.org/10.1093/imanum/drx029
3. Ascher, U.M., Ruuth, S.J., Spiteri, R.J.: Implicit-explicit Runge-Kutta methods for time-dependent partial differential equations. Appl. Numer. Math. **25**(2), 151–167 (1997). https://doi.org/10.1016/S0168-9274(97)00056-1
4. Barabási, A.L., Albert, R.: Emergence of scaling in random networks. Science **286**(5439), 509–512 (1999)
5. Berkolaiko, G.: An elementary introduction to quantum graphs. Geom. Comput. Spectr. Theory **700**, 41–72 (2017)
6. Berkolaiko, G., Kuchment, P.: Introduction to Quantum Graphs. Math. Surv. Monogr. American Mathematical Society, Providence (2013)
7. Besse, C., Duboscq, R., Le Coz, S.: Numerical simulations on nonlinear quantum graphs with the GraFiDi library. SMAI J. Comput. Math **8**, 1–47 (2022). https://doi.org/10.5802/smai-jcm.78
8. Bischof, G.N., Jessen, F., Fliessbach, K., Dronse, J., Hammes, J., Neumaier, B., Onur, O., Fink, G.R., Kukolja, J., Drzezga, A., van Eimeren, T.: Impact of tau and amyloid burden on glucose metabolism in Alzheimer's disease. Ann. Clin. Transl. Neurol. **3**(12), 934–939 (2016). https://doi.org/10.1002/acn3.339
9. Brenner, S.C., Scott, L.R.: The Mathematical Theory of Finite Element Methods, 3rd edn. Springer, New York (2008)
10. Brio, M., Caputo, J.G., Kravitz, H.: Spectral solutions of PDEs on networks. Appl. Numer. Math. **172**, 99–117 (2022). https://doi.org/10.1016/j.apnum.2021.09.021
11. Brouwer, A.E., Haemers, W.H.: Spectra of Graphs. Springer, New York (2011)
12. Canuto, C., Hussaini, M.Y., Quarteroni, A., Zang, T.A.: Spectral Methods: Fundamentals in Single Domains. Springer Berlin, Heidelberg (2007)
13. Cattaneo, C.: The spectrum of the continuous laplacian on a graph. Monatsh. Math. **124**(3), 215–235 (1997). https://doi.org/10.1007/BF01298245
14. Chou, I.h.: Read my mind. Nat. Phys. **4**(1), S17–S17 (2008). https://doi.org/10.1038/nphys874
15. Chung, F.R.K.: Spectral Graph Theory. CBMS 92. American Mathematical Society, Providence (1997)

A. Weller, *Numerical Methods for Metric Graphs*, Lecture Notes in Mathematics 2382, https://doi.org/10.1007/978-3-032-05011-3

16. Cope, T., Rittman, T., Borchert, R., Jones, P., Vatansever, D., Allinson, K., Passamonti, L., Vazquez Rodriguez, P., Bevan-Jones, W., O'Brien, J., Rowe, J.: Tau burden and the functional connectome in Alzheimer's disease and progressive supranuclear palsy. Brain **141**(2), 550–567 (2018). https://doi.org/10.1093/brain/awx347
17. Dröge, C.S., Weller, A.: Computation of quantum graph spectra. Conference Poster at Rhein-Ruhr Workshop (2023)
18. Dröge, C.S., Weller, A.: Numerical computation of quantum graph spectra. arXiv preprint. arXiv:2312.09116 (2023)
19. Drzezga, A.: The network degeneration hypothesis: spread of neurodegenerative patterns along neuronal brain networks. J. Nucl. Med. **59**(11), 1645–1648 (2018). https://doi.org/10.2967/jnumed.117.206300
20. Eisenberg, D., Jucker, M.: The amyloid state of proteins in human diseases. Cell **148**(6), 1188–1203 (2012). https://doi.org/10.1016/j.cell.2012.02.022
21. Erdős, P., Rényi, A.: On the evolution of random graphs. Publ. Math. Inst. Hung. Acad. Sci. **5**(1), 17–60 (1960)
22. Evans, L.C.: Partial Differential Equations. Graduate Studies in Mathematics. American Mathematical Society, Providence (2000)
23. Filon, L.N.G.: On a quadrature formula for trigonometric integrals. Proc. R. Soc. Edinb. **49**, 38–47 (1930)
24. Goedert, M., Eisenberg, D.S., Crowther, R.A.: Propagation of tau aggregates and neurodegeneration. Annu. Rev. Neurosci. **40**, 189–210 (2017). https://doi.org/10.1146/annurev-neuro-072116-031153
25. Goodman, R.H., Conte, G., Marzuola, J.L.: Qglab: a matlab package for computations on quantum graphs. SIAM J. Sci. Comput. **47**(2), B428–B453 (2025). https://doi.org/10.1137/23M1627729
26. Güttel, S., Tisseur, F.: The nonlinear eigenvalue problem. Acta Numer. **26**, 1–94 (2017). https://doi.org/10.1017/S0962492917000034
27. Hackbusch, W.: Parabolic multi-grid methods. In: Comput. Methods Appl. Sci. Eng. VI. North-Holland, Amsterdam (1984)
28. Hackbusch, W.: Iterative Solution of Large Sparse Systems of Equations. Appl. Math. Sci. Springer International Publishing, New York (2016)
29. Hochbruck, M., Ostermann, A.: Explicit exponential Runge-Kutta methods for semilinear parabolic problems. SIAM J. Numer. Anal. **43**(3), 1069–1090 (2006)
30. Hoenig, M.C., Bischof, G.N., Seemiller, J., Hammes, J., Kukolja, J., Onur, Ö.A., Jessen, F., Fliessbach, K., Neumaier, B., Fink, G.R., van Eimeren, T., Drzezga, A.: Networks of tau distribution in Alzheimer's disease. Brain **141**(2), 568–581 (2018). https://doi.org/10.1093/brain/awx353
31. Hoenig, M.C., Weller, A., Doering, E., Bischof, G.N., Drzezga, A., van Eimeren, T.: Functional connectivity serves as transmitter for the remote effects of regional amyloid on tau pathology spread. In: Nuklearmedizin 2023, **62**(02), 83 (2023)
32. Hofmann, M.: Spectral theory, clustering problems and differential equations on metric graphs. Ph.D. thesis, Universidade de Lisboa (Portugal) (2021)
33. Hu, G.Y., O'Connel, R.F.: Analytical inversion of symmetric tridiagonal matrices. J. Phys. A Math. Gen. **29**(7), 1511 (1996)
34. Kennedy, J.B., Kurasov, P., Malenová, G., Mugnolo, D.: On the spectral gap of a quantum graph. Ann. Henri Poincaré **17**(9), 2439–2473 (2016). https://doi.org/10.1007/s00023-016-0460-2
35. Kuchment, P.: Quantum graphs: I. Some basic structures. Waves Random Media **14**(1), S107 (2003). https://doi.org/10.1088/0959-7174/14/1/014
36. Kunoth, A., Weller, A., Yilmaz, T.: A computational brainsphere model for the simulation of Alzheimer's disease. Snapshot of modern mathematics from Oberwolfach (submitted, April 2024)
37. Kurasov, P., Stenberg, F.: On the inverse scattering problem on branching graphs. J. Phys. A Math. Gen. **35**(1), 101 (2002). https://doi.org/10.1088/0305-4470/35/1/309

38. Lancaster, P.: Algorithms for lambda-matrices. Numer. Math. **6**(1), 388–394 (1964). https://doi.org/10.1007/BF01386088
39. Lancaster, P.: Some numerical methods for lambda-matrices. In: Lambda-Matrices and Vibrating Systems. Int. Ser. Monogr. Pure Appl. Math., vol. 94, pp. 75–99. Pergamon (1966). https://doi.org/10.1016/B978-0-08-011664-8.50009-2
40. Le Bihan, D., Mangin, J.F., Poupon, C., Clark, C.A., Pappata, S., Molko, N., Chabriat, H.: Diffusion tensor imaging: concepts and applications. J. Magn. Reson. Imaging **13**(4), 534–546 (2001)
41. Mugnolo, D.: Semigroup Methods for Evolution Equations on Networks. Underst. Complex Syst. Springer International Publishing, Cham (2014)
42. Pankrashkin, K.: Spectra of Schrödinger operators on equilateral quantum graphs. Lett. Math. Phys. **77**(2), 139–154 (2006). https://doi.org/10.1007/s11005-006-0088-0
43. Pauling, L.: The diamagnetic anisotropy of aromatic molecules. J. Chem. Phys. **4**(10), 673–677 (1936). https://doi.org/10.1063/1.1749766
44. Post, O.: Spectral analysis of metric graphs and related spaces. In: Limits Graphs Group Theory Comput. Sci., pp. 109–140 (2008)
45. Raj, A., Kuceyeski, A., Weiner, M.: A network diffusion model of disease progression in dementia. Neuron **73**(6), 1204–1215 (2012). https://doi.org/10.1016/j.neuron.2011.12.040
46. Rohr, C.S., Okon-Singer, H., Craddock, R.C., Villringer, A., Margulies, D.S.: Affect and the brain's functional organization: a resting-state connectivity approach. PLoS One **8**(7), 1–13 (2013). https://doi.org/10.1371/journal.pone.0068015
47. Schäfer, A., Mormino, E.C., Kuhl, E.: Network diffusion modeling explains longitudinal tau PET data. Front. Neurosci. **14**, 566876 (2020). https://doi.org/10.3389/fnins.2020.566876
48. Schanz, H.: A relation between the bond scattering matrix and the spectral counting function for quantum graphs. Contemp. Math. **415**, 269 (2006)
49. Stoll, M., Winkler, M.: Optimal Dirichlet control of partial differential equations on networks. Electron. Trans. Numer. Anal. **54**, 392–419 (2021)
50. Thomée, V.: Galerkin Finite Element Methods for Parabolic Problems. No. 25 in Comput. Math. Ser. Springer Berlin, Heidelberg (1997)
51. Vogel, J.W., Iturria-Medina, Y., Strandberg, O.T., Smith, R., Levitis, E., Evans, A.C., Hansson, O.: Spread of pathological tau proteins through communicating neurons in human Alzheimer's disease. Nat. Commun. **11**(1), 2612 (2020)
52. von Below, J.: Diffusion und Reaktion auf Netzwerken. Ph.D. thesis, Univ. of Tubingen (1984)
53. von Below, J.: A characteristic equation associated to an eigenvalue problem on c2-networks. Linear Algebra Appl. **71**, 309–325 (1985). https://doi.org/10.1016/0024-3795(85)90258-7
54. Weller, A.: MeGraPDE.jl - a julia package for PDEs on metric graphs (2023). https://github.com/AnnaWeller/MeGraPDE.jl
55. Weller, A., Bischof, G.N., Schlüter, P., Richter, N., Neumaier, B., Kukolja, J., Kunoth, A., Shao, Y., van Eimeren, T., Drzezga, A.: Graph theoretical analysis of tau burden and the functional connectome in Alzheimer's disease (2018). Poster presented at the 2nd Molecular Imaging of Neurodegeneration in Cologne Symposium
56. Weller, A., Bischof, G.N., Schlüter, P., Richter, N., Dronse, J., Onur, O., Neumaier, B., Kukolja, J., Langen, K.J., Fink, G., Kunoth, A., Shao, Y., van Eimeren, T., Drzezga, A.: Finding new communities: a principle of neuronal network reorganization in Alzheimer's disease. Brain Connect. **11**(3), 225–238 (2021). https://doi.org/10.1089/brain.2020.0889

Index

A. Weller, *Numerical Methods for Metric Graphs*, Lecture Notes in Mathematics
2382, https://doi.org/10.1007/978-3-032-05011-3

LECTURE NOTES IN MATHEMATICS Springer

Editors in Chief: J.-M. Morel, B. Teissier;

Editorial Policy

1. Lecture Notes aim to report new developments in all areas of mathematics and their applications – quickly, informally and at a high level. Mathematical texts analysing new developments in modelling and numerical simulation are welcome.

   Manuscripts should be reasonably self-contained and rounded off. Thus they may, and often will, present not only results of the author but also related work by other people. They may be based on specialised lecture courses. Furthermore, the manuscripts should provide sufficient motivation, examples and applications. This clearly distinguishes Lecture Notes from journal articles or technical reports which normally are very concise. Articles intended for a journal but too long to be accepted by most journals, usually do not have this "lecture notes" character. For similar reasons it is unusual for doctoral theses to be accepted for the Lecture Notes series, though habilitation theses may be appropriate.

2. Besides monographs, multi-author manuscripts resulting from SUMMER SCHOOLS or similar INTENSIVE COURSES are welcome, provided their objective was held to present an active mathematical topic to an audience at the beginning or intermediate graduate level (a list of participants should be provided).

   The resulting manuscript should not be just a collection of course notes, but should require advance planning and coordination among the main lecturers. The subject matter should dictate the structure of the book. This structure should be motivated and explained in a scientific introduction, and the notation, references, index and formulation of results should be, if possible, unified by the editors. Each contribution should have an abstract and an introduction referring to the other contributions. In other words, more preparatory work must go into a multi-authored volume than simply assembling a disparate collection of papers, communicated at the event.

3. Manuscripts should be submitted either online at www.editorialmanager.com/lnm to Springer's mathematics editorial in Heidelberg, or electronically to one of the series editors. Authors should be aware that incomplete or insufficiently close-to-final manuscripts almost always result in longer refereeing times and nevertheless unclear referees' recommendations, making further refereeing of a final draft necessary. The strict minimum amount of material that will be considered should include a detailed outline describing the planned contents of each chapter, a bibliography and several sample chapters. Parallel submission of a manuscript to another publisher while under consideration for LNM is not acceptable and can lead to rejection.

4. In general, **monographs** will be sent out to at least 2 external referees for evaluation.

   A final decision to publish can be made only on the basis of the complete manuscript, however a refereeing process leading to a preliminary decision can be based on a pre-final or incomplete manuscript.

   Volume Editors of **multi-author works** are expected to arrange for the refereeing, to the usual scientific standards, of the individual contributions. If the resulting reports can be forwarded to the LNM Editorial Board, this is very helpful. If no reports are forwarded or if other questions remain unclear in respect of homogeneity etc, the series editors may wish to consult external referees for an overall evaluation of the volume.

5. Manuscripts should in general be submitted in English. Final manuscripts should contain at least 100 pages of mathematical text and should always include
   - a table of contents;
   - an informative introduction, with adequate motivation and perhaps some historical remarks: it should be accessible to a reader not intimately familiar with the topic treated;
   - a subject index: as a rule this is genuinely helpful for the reader.
   - For evaluation purposes, manuscripts should be submitted as pdf files.

6. Careful preparation of the manuscripts will help keep production time short besides ensuring satisfactory appearance of the finished book in print and online. After acceptance of the manuscript authors will be asked to prepare the final LaTeX source files (see LaTeX templates online: https://www.springer.com/gb/authors-editors/book-authors-editors/manuscriptpreparation/5636) plus the corresponding pdf- or zipped ps-file. The LaTeX source files are essential for producing the full-text online version of the book, see http://link.springer.com/bookseries/304 for the existing online volumes of LNM). The technical production of a Lecture Notes volume takes approximately 12 weeks. Additional instructions, if necessary, are available on request from lnm@springer.com.

7. Authors receive a total of 30 free copies of their volume and free access to their book on SpringerLink, but no royalties. They are entitled to a discount of 33.3 % on the price of Springer books purchased for their personal use, if ordering directly from Springer.

8. Commitment to publish is made by a *Publishing Agreement*; contributing authors of multiauthor books are requested to sign a *Consent to Publish form*. Springer-Verlag registers the copyright for each volume. Authors are free to reuse material contained in their LNM volumes in later publications: a brief written (or e-mail) request for formal permission is sufficient.

Addresses:
Professor Jean-Michel Morel, CMLA, École Normale Supérieure de Cachan, France
E-mail: moreljeanmichel@gmail.com

Professor Bernard Teissier, Equipe Géométrie et Dynamique,
Institut de Mathématiques de Jussieu – Paris Rive Gauche, Paris, France
E-mail: bernard.teissier@imj-prg.fr

Springer: Ute McCrory, Mathematics, Heidelberg, Germany,
E-mail: lnm@springer.com

MIX
Papier aus verantwortungsvollen Quellen
Paper from responsible sources
FSC® C105338

If you have any concerns about our products,
you can contact us on
ProductSafety@springernature.com

In case Publisher is established outside the EU,
the EU authorized representative is:
Springer Nature Customer Service Center GmbH
Europaplatz 3, 69115 Heidelberg, Germany

Printed by Libri Plureos GmbH
in Hamburg, Germany